To our cousin Peter and Bonnie Marsh;
with deepest affection and best wishes –
Jan P. Bailly
23 October 2007

One Time Fits All

N. C. Wyeth, *illustration from* Robinson Crusoe, *Cosmopolitan Book, 1920. Original in the Robinson Crusoe Collection, the Wilmington Institute Library, Wilmington, Delaware.*

One Time Fits All

The Campaigns for Global Uniformity

Ian R. Bartky

STANFORD UNIVERSITY PRESS

STANFORD, CALIFORNIA

2007

Stanford University Press
Stanford, California
© 2007 by the Board of Trustees of the Leland
Stanford Junior University.
All rights reserved.

Printed in the United States of America on acid-free,
archival-quality paper

Library of Congress Cataloging-in-Publication Data

Bartky, Ian R.
  One time fits all : the campaigns for global
uniformity / Ian R. Bartky.
      p. cm.
  Includes bibliographical references and index.
  ISBN 978-0-8047-5642-6 (cloth : alk. paper)
  1. Time--Systems and standards. I. Title.

QB223.B37 2007
389'.17--dc22   2007015373

Typeset by Bruce Lundquist in 10/12 Sabon

In memory of mentors
Walter Bartky, University of Chicago,
and
William F. Giauque, University of California, Berkeley,
whose guidance persists;
and to
Elizabeth Hodgins Bartky,
my steadfast partner in this endeavor.

Contents

Contents

Illustrations

Figures

Tables

Preface

While completing *Selling the True Time* (Stanford University Press, 2000), my study of public timekeeping in America in the nineteenth century, I was struck by how easy it is today to know the exact time to within a second or two. Accurate time is available everywhere: on quartz watches, cell phones, GPS receivers, radio-controlled wall clocks, personal computers, and home entertainment components that reset their displays automatically after power failures and even adjust for the twice-yearly shift between Standard Time and Daylight Saving Time.

These modern time givers acquire and maintain the correct time thanks to an extensive set of accords among the world's time providers working at physics laboratories and astronomical observatories in a host of countries. Their technical accords are virtually transparent to the general public. However, how the consumer devices adjust to the other divisions of civil time is also dependent on agreements, ones that directly affect people. This book explores how these latter conventions were reached and how they have evolved.

In my earlier study I described many of the public timekeeping conventions adopted in the United States and told how they came about. I deliberately omitted any discussion of twentieth-century timekeeping and, by focusing on America, excluded significant nineteenth-century events that took place elsewhere. This new book concentrates on three time-related subjects that have continuing importance for travelers and ordinary citizens in every country: the International Date Line, the system of standard time zones that encompasses the globe, and Daylight Saving Time.

These subjects are all linked to international acceptance of Greenwich as the prime meridian. This acceptance was achieved only slowly and involved more than fifty years of intense discussion; the matter was eventually resolved between 1870 and 1925. This latter date also marks the culmination of efforts aimed at unifying the timekeeping of specialists by linking their calculations and tables to Greenwich civil time.

Although printed sources cover certain facets of the areas identified here, they are sadly incomplete. For example, no history of the International Date Line exists that accurately describes the tremendous alterations in its sweep across the Pacific Ocean over the last two centuries, or even how its location was eventually fixed. Similarly, there has been no detailed study of how the world's time zones came into being.

Web-based sources that deal with these areas are notoriously unreliable. Indeed, much of what is being distributed electronically as fact is incomplete and often dead wrong. For example, every site I examined stated that the International Meridian Conference of 1884 resolved uniformity issues and led directly to many of our current conventions regarding the public's time. My conclusion from examining original records and correspondence is diametrically opposite: the conference was a failure. To set the record straight I have made every effort to ensure that every statement offered here can be traced back to original sources.

My previous book included no visual record of those who contributed significantly to uniformity in American timekeeping. This omission was noted by a good friend who reviewed the work and, later, by one of my colleagues, who urged me to include portraits in this new study. I have done so, locating photographs and engravings created, for the most part, close to the time when key characters were most active. In my view, it is important to show them as they were then, for late-in-life images often give a false impression.

Many of those named and pictured, while well-known in their own countries, are virtually unknown elsewhere. One reason for this is that historians are often parochial, especially when they focus on their own country. For example, no previous English-language writer has done more than hint at how much French activities during the first decades of the twentieth century contributed to global efforts at unifying time. Realizing their importance has led me to devote an entire chapter to them in this study.

For some time now it has seemed that any public interest in Daylight Saving Time was essentially historical and, further, that the boundaries of all U.S. time zones were now fixed. Imagine my surprise when last year not one but two books on the subject appeared: Michael Downing's *Spring Forward* (Shoemaker & Hoard, 2005) and David Prerau's *Seize the Daylight* (Thunder's Mouth Press, 2005). Both of these popularly written accounts, which have vastly different views regarding Daylight Saving Time's impact, are filled primarily with anecdotes from the last eighty years. In contrast, I chose to end my story in the 1920s so as to focus on the basic issues underlying more than a century of efforts directed at changing America's public's timekeeping.

However, Indiana's recent decision to begin observing Daylight Saving Time everywhere (starting in 2006) and the passage of the Energy Policy Act of 2005, which mandates extending Daylight Saving Time in the United States beginning in 2007, caused me to add to the epilogue a discussion of some of the trade-offs that affect the public—trade-offs articulated in 1976 in testimony before Congress.

My first professional encounter with Daylight Saving Time happened over thirty years ago. Its public impact led me to undertake historical studies of timekeeping and of efforts to make time uniform throughout the United States and the world.

Acknowledgments

Writing an extended history is a team effort, and my obligations to the many individuals who have contributed to this endeavor span two continents and the South Sea Islands. Some I have never met, yet, thanks to e-mail, I have been able to gain their assistance and acquire their knowledge. Further, because I live near Washington, D.C., I have been able to draw upon the magnificent resources of several national libraries and archives.

The U.S. Naval Observatory's librarians Brenda Corbin and Gregory Shelton not only guided me through their exceptional book and photographic collections, but also located countless items at other libraries and helped me gain access to them. Its public affairs officer, Geoff Chester, located Navy organizations whose names had been altered so many times that the task of keeping track of them in my narrative often seemed hopeless, and also supervised electronic scans for my specific needs. Steven Dick, now NASA's chief historian, provided useful citations to Naval Observatory activities associated with the controversy over the astronomical day.

Without the tireless assistance of the Library of Congress staff, I could never have completed this study. Heading the list is that superb and gifted enabler Constance Carter, of the Science and Technology Division, who, with Margaret Clifton, located in the library's stacks a veritable treasure trove of poorly cited source documents. The Law Library's Clare Feikert zeroed in on several obscure British government sources, thereby greatly enhancing our research efforts in London. John Hébert and Ed Redmond, respectively chief of and map specialist in the library's Geography and Map Division, provided much-appreciated support as I began my quest for historical maps and charts, as did the acquisitions specialist Jim Flatness; and our friend Ralph Ehrenberg, former division chief in charge of this fabulous resource, photographed maps in the collection from which I made tracings. Curator Elena Millie in the Prints and Photographs Division helped me track down Daylight Saving Time propaganda posters, and these, too, contribute significantly to my story of twentieth-century public time. And in the Rare Book and Special Collections Division, the reference librarian Clark Evans and his Reading Room staff located numerous accounts of the early circumnavigators, which made it possible for me to provide a history of the changes in daily reckoning among explorers.

Staff at Washington-area technical libraries were most generous with their time. At NOAA's Central Library, Albert E. "Skip" Theberge, Jr., helped me locate images in the collection, and the librarian Mary Lou Cumberpatch guided me through the intricacies of the separate card catalogs in my search for journals and documents, assisting me, as did Caroline Woods, whenever I went astray. At the U.S. Geological Survey, the research librarian Brenda Graff assisted us both before and during several visits, thereby making our search efforts much more efficient. Staff at NIST's library, the Navy Department Library, and the E. B. Williams Law Library at the Georgetown University Law Center helped us locate specialized journals, historical documents, and congressional materials not available elsewhere.

At the National Archives and Records Administration, Sharon Thibodeau continued her decades of outstanding assistance by guiding me through the vast files of U.S. Hydrographic Office maps and State Department correspondence. The archivists Richard Peuser and Rebecca Livingston were instrumental in making my search of nineteenth-century Navy ship logs profitable.

In Ohio David P. Whelan, director of the Cincinnati Law Library Association, and Linda Bailey, librarian at the Cincinnati Historical Society Library, very kindly located the several documents associated with the Queen City of the West's early advanced-time ordinance.

Some collections on the West Coast were made accessible to me via e-mail, which saved me considerable time and expense. These included the Mary Lea Shane Archives of Lick Observatory, whose curator, Dorothy Schaumberg, assisted by Cheryl Dandridge, located photographs of astronomers within my range of interest. As often before, I once again found her skills impressive. Thanks to William Roberts and an excellent finding aid, the Bancroft Library at the University of California, Berkeley (a favorite resource since my graduate student days), supplied copies of George Davidson's correspondence. I am also grateful to Bancroft's Erica Nordmeier, who made sure a key map was scanned before the library's seismic retrofit made its storage location temporarily inaccessible. At the Hoover Institution Archives, Stanford University, Carol Leadenham located World War I German posters publicizing Summer Time, and

Heather Wagner made sure that copies reached me in time to be included in this book. Near the end of our research, Stephen Watrous, emeritus professor of history at Sonoma State University, California, pointed me to the map of the Fort Ross settlement that is included in this work; and Lyn Kalani, executive director of the Fort Ross Interpretive Association, provided other valuable research materials.

Additional support for the research effort came from New England institutions. Our sojourn at the Phillips Library in Salem, Massachusetts, where my wife and I reviewed a large number of nineteenth-century merchant ship logs under the guidance of the librarian Britta Karlberg, was enhanced by the kindnesses shown us by the management of the Peabody Essex Museum, arranged by Charity Galbreath, to whom we are most grateful. At the G. W. Blunt White Library, Mystic Seaport, Connecticut, the research manager Wendy Schnur piloted two landlubbers efficiently and safely through the shoals of false starts; now I almost know the precise meaning of the word "ship." Wendy Anthony, of Special Collections, Lucy Scribner Library, Skidmore College, Saratoga Springs, New York, helped establish the date of a likeness of the time-reform champion Charles Dowd. Beinecke Library staff and, once again, William R. Massa, Jr., of Sterling Memorial Library, speeded the research effort at Yale. Maria McEachern at the John Wolbach Library, Harvard-Smithsonian Center for Astrophysics, supplied copies of pamphlets unavailable elsewhere and guided me to staff at Harvard's Widener Library for additional items. David Burdash, director of the Wilmington Institute, Delaware, gave me permission to use one of N. C. Wyeth's paintings of Robinson Crusoe in the collection.

A significant part of our research effort took place in Ottawa at the Library and Archives Canada. The reference archivist Sophie Tellier made many arrangements for us, giving us an extremely efficient way to review the archives' large collection of Sandford Fleming materials housed there, and the librarian Tom Tytor located copies of all the Canadian engineer's first pamphlets. During our Ottawa visits, Randall Brooks, of the Canada Science and Technology Museum, extended many kindnesses and made several suggestions regarding my quest. The civil engineer Mark Andrews opened his collection of historical engineering publications and sent me several digital images.

A number of European experts, too, rendered invaluable assistance. Among them was Carlos Nélson Lopes da Costa, technical director of Portugal's Instituto Hidrográfico, who provided documentation for several of his organization's charts, as did Raymond Jehan of France's Service Hydrographique et Océanographique de la Marine. The astronomer Jean-Pierre De Cuyper, Royal Observatory of Belgium, with Arnold Fremout, head of the Hydrographic Service of Flanders, and his colleague Johan Verstraeten, supplied details on the country's early hydrographic charts. I am particularly grateful to Captain Hugo Gorziglia, director of the International Hydrographic Bureau, Monaco, who gave me permission to use an image of the French hydrographer Joseph Renaud. And long-time Washington friend and retired District Court of Appeals judge John Steadman translated documents associated with Spanish hydrography.

The historian of science Henning Schmidgen, currently with the Faculty of Arts and Sciences, Harvard University, guided my search for contemporary European images, and the archive director Erna Lämmel of the German Academy of Naturalists "Leopoldina" in Halle very generously supplied a photograph of the Austrian geodesist Robert Schram. I am also grateful to BIPM's director emeritus, Terry Quinn, for suggesting sources with regard to Great Britain and the Convention of the Metre, and to the librarian Danièle Le Coz, who helped in dating a photograph of Adolphe Hirsch. Guy Coronio, editor of the Presses de l'École nationale des ponts et chaussées, very kindly granted permission to use an image of Charles Lallemand from one of its French geodesy publications.

Our obligations to individuals and organizations in the United Kingdom are many. First, I acknowledge with gratitude the award of a Caird Library Short-Term Research Fellowship by the National Maritime Museum, Greenwich. Without it, some of the research we undertook during 2002 could not have been accomplished. Janet Norton was instrumental in making us feel part of the museum community, as was Sally Archer a few years later. We are also indebted to Maria Blyzinski.

We made several research trips to the United Kingdom while this work was in progress. Every time, my colleagues at the museum's Royal Greenwich Observatory—Jonathan Betts, Gloria Clifton, and David Rooney—extended themselves, locating documents, suggesting areas to consider, and listening to the tales of my progress. Without their enthusiastic support, I could not have proceeded so rapidly in this endeavor.

I am grateful to the Council for World Mission, London, for permission to examine the archives of the London Missionary Society, held at SOAS, University of London. The chief executive, Roland Jackson, very kindly gave me similar permission to examine the BAAS archives, housed at Oxford University's Bodleian Library; the Bodleian's Oliver House made all arrangements prior to our arrival, which saved us countless hours. Neil Brown at the Science Museum, London, steered me to the museum's holdings of the annual reports of the Science and Art Department, documents crucial to understanding the British government's role in advancing time uniformity. Adrian Webb at UKHO, Taunton, located historical Admiralty charts and supplied additional information regarding them; Sharon Nichol and Lucy Hollands provided further information. I wish particularly to thank Francis Herbert, Curator of Maps, Royal Geographical Society, London, who located important correspondence in the archives; and the archivist Sara Strong, who made them available. Finally, library staff at Cambridge University and the Royal Astronomical Society were extremely helpful.

In addition I acknowledge with pleasure permission to quote from a copy of Captain Philip Pipon's 1814 narrative in the Sir Joseph Banks Electronic Archives, State Library of New South Wales, Australia.

On reviewing the research efforts undertaken over the past years, it became clear that three people, more than any others, vastly increased my understanding of the subject areas. Again and again I would call on them when I needed guidance; and each time they were unstinting in their response. These experts are the retired U.K. Ordnance Survey officer and map historian Ian Mumford; the map archivist and historian Andrew Cook, India Office Records, British Library; and the former career diplomat for New Zealand Rhys Richards, a maritime historian specializing in Pacific Ocean history prior to 1850.

On reaching the writing and editing stages, my friend Lane Jennings, a poet, author, and editor, carefully read every chapter and endnote, suggesting changes in emphasis and wording to strengthen and clarify the text. At every turn, his remarks were invaluable.

Thanks to Norris Pope, Director of Scholarly Publications, my interactions with Stanford University Press have been enormously satisfying. Throughout so many years, Norris encouraged me, reacted to my enthusiasms with positive and stimulating suggestions, and on several occasions guided me with regard to intellectual property issues. I count him as a friend.

At the press, the assistant editor Emily-Jane Cohen oversaw the nearly infinite details that transform a computer disc and its associated bundles of text and illustrations from manuscript to a published work—labors that fill me with awe. The production editor Carolyn Brown resolved myriad critical details throughout the process; it has been extremely pleasant to interact with such an enthusiastic and careful expert. I am most grateful to all who aided me.

Abbreviations

AAAS   American Association for the Advancement of Science
ADM   Admiralty
AMS   American Metrological Society
ASCE   American Society of Civil Engineers
BAA   British Astronomical Association
BAAS   British Association for the Advancement of Science
BIH   International Bureau of Time
BIPM   International Bureau of Weights and Measures
CDT   Central Daylight Time
CGPM   General Conference on Weights and Measures
CIA   Central Intelligence Agency
CIPM   International Committee on Weights and Measures
*Comptes rendus   Comptes rendus hebdomadaires des séances de
      l'Académie des Sciences*
Cong.   Congress
Cong. Rec.   Congressional Record
CST   Central Standard Time
DDST   Double Daylight Saving Time
DOT   Department of Transportation
DST   Daylight Saving Time
EDT   Eastern Daylight Time
EST   Eastern Standard Time
GPO   Government Printing Office

GPS Global Positioning System

GTC General Time Convention

H. Ex. Doc. House Executive Document

HIA Hoover Institution Archives, Stanford University

H. J. Res. House Joint Resolution

HMS His (Her) Majesty's Ship

H.O. U.S. Hydrographic Office

H.R. House Resolution

H. Rept. House Report

ICC Interstate Commerce Commission

IGA International Geodetic Association (*Europäische Gradmessung*)

IGC International Geographical Congress

IHB International Hydrographic Bureau, Monaco

IHO International Hydrographic Organization

LC Library of Congress

LMS London Missionary Society

Misc. Doc. Miscellaneous Document

NARA National Archives and Records Administration

NAS National Academy of Sciences

NBS National Bureau of Standards (now NIST)

NIST National Institute of Standards and Technology

NOAA National Oceanic and Atmospheric Administration

N.S. New Style, Gregorian calendar reckoning

Ohio Education *The Ohio Educational Monthly and The National Teacher*

O.S. Old Style, Julian calendar reckoning

PAMS *Proceedings of the American Metrological Society*

P.L. Public Law

RAS Royal Astronomical Society (London)

RG Record Group

R.G.S. Royal Geographical Society (London)

S. Senate bill

School Bulletin   The School Bulletin and New York State Educational Journal

S. Ex. Doc.   Senate Executive Document

SHOM   *Service Hydrographique et Océanographique de la Marine*

S.J.   Society of Jesus

S. J. Res.   Senate Joint Resolution

SOAS   School of Oriental and African Studies, University of London

S. Rept. No.   Senate Report Number

U.K.   United Kingdom of Great Britain and (Northern) Ireland

UKHO   United Kingdom Hydrographic Office

USC&GS   U.S. Coast & Geodetic Survey

USGS   U.S. Geological Survey

USHO   U.S. Hydrographic Office

USN   U.S. Navy

USNO   U.S. Naval Observatory

UTC   Universal Coordinated Time

YRDST   Year-round Daylight Saving Time

One Time Fits All

Introduction

I found I had been there a year, so I divided it into weeks, and set
apart every seventh day for a Sabbath. . . . [B]ut as for an exact reckoning of days, after I had once lost it, I could never recover it again.
—Daniel Defoe, *Robinson Crusoe*, 1719

The integrative theme in this work is the drive for uniformity as it affects the public. Without stable systems of consistent measurement of all
kinds, the vast majority of our daily transactions would be problematic
at best. For example, absent carefully crafted agreements regarding public time among regions and countries, a passing stranger's query "What
time do you have?" would require a ponderous response. And without a
common way to locate places on maps and charts, we would always be
making back-and-forth conversions while estimating distances—producing errors during the process.

Time—mean solar time as displayed on a clock—and longitude are
linked. On average, one complete rotation of the Earth about its axis
takes twenty-four hours, and the globe's circumference is divided uniformly into 360 degrees. Thus one hour of mean time corresponds to
15 degrees of longitude, and in the same way 1 degree of longitude corresponds to four minutes in time.[1] In addition, placing a network (graticule) of parallels and meridians encompassing the surface of the earth
in order to designate locations has been employed for centuries, with
parallels of latitude enumerated as (0–90) degrees north or south of the
equator. Unlike latitude, however, there is no obvious start point for
the meridian lines that specify east–west positions. Throughout most of
history, map and chart makers chose arbitrary beginning points (often
a landmark in their own locality) for the initial meridian of longitude.
A few hundred years ago it became apparent that adopting a common

initial longitude offered advantages to ocean navigators and geographers, but not until a century and a half ago was this feasible.

Like the suggestions for unifying longitudes, proposals for unifying and simplifying the public's reckoning of time came from experts anxious to resolve some particular discrepancy or ambiguity. Efforts to change people's habits, however, generally meet stiff resistance, for most individuals are quite comfortable with their accustomed ways and see no reason to change. Consequently, most reforms in time reckoning have come only after decades, sometimes centuries, of intense discussions and negotiations.

Perhaps the most famous alteration in time reckoning was the replacement of the Julian calendar by the Gregorian one. In 46 BC, Julius Caesar reformed the Roman calendar, then badly out of synchrony with the seasons. His calendar was a solar one having 365 days in a year with an additional day inserted once every four years, giving an average calendar year of 365¼ days. Since this period was slightly longer than the length of an astronomical year—the required interval for the earth to circle the sun and return to the same place in its orbit—astronomically based events began occurring earlier and earlier according to the Julian calendar. The discrepancy between the two was small, a shift of one day in 128 years, but it grew as the centuries accumulated. By 1300 the vernal (spring) equinox, a defining point in the earth's orbit, fell on 11 March, ten days early. Moreover, full moons now came three days earlier by the Julian calendar. As a result, calculations for Easter Day, which use both calendar values, were wrong.

In 1582, after centuries of suggestions for reforming the calendar, several failed attempts to do so, and a difference now amounting to ten days, Pope Gregory XIII accepted the report of a commission he had established to consider the various proposals. On 25 February he signed a papal bull to put the changes into effect. Ten days were dropped, so that 15 October came just after 4 October, and the vernal equinox was returned to 21 March. In addition, to ensure that the calendar year and the astronomical one now coincided, the Julian calendar's leap-year rule was altered: Every four years a day would continue to be inserted, but with the exception that only those century years divisible by four hundred (i.e., 1600, 2000, etc.) would contain the extra day—that is, 1700, 1800, 1900, 2100, and so on are not leap years.[2]

The switch in calendars took place in Catholic Europe starting in October 1582 but as late as 1585 in Spain's overseas possessions.[3] England and her colonies did not adopt the Gregorian calendar until September 1752, nearly two hundred years after it was first instituted, and its adoption by countries such as China, Soviet Union, Greece, and Turkey did not occur until the twentieth century. Even today, the Gregorian reckoning is still not used by all practicing Christians, nor by Muslims and Jews, who are guided by their own religious calendars. Despite these exceptions as well as a host of proposals from reformers anxious to improve the "regularity" of calendar months, the Gregorian calendar is the calendar for commerce everywhere and for nonreligious time reckoning. It is likely to remain so.[4]

The day itself was also partitioned early on, with the Egyptians dividing it into twenty-four parts, twelve for the nights and twelve for the daylight periods. Slowly, and then more rapidly with the fourteenth-century invention of the clock and subsequent improvements in accuracy, these twenty-four parts, long of unequal duration according to the seasons, became equal. Time by the clock, with hours of equal duration, became the norm, and the average of the length of a year of solar days—the mean solar day—became the standard.

Much of our story takes place during the era when, for public activities, time by the clock was customary, with its divisions of the hour into minutes and seconds and even small fractions of a second. Every place was keeping its own civil time: mean solar time, with noon when the (mean) sun crossed the locale's meridian. These differences in local time varied systematically and in a simple-to-calculate way: the civil time of any place to the east of one's locale was later, the civil time of any place to the west was earlier.

Our story begins in the sixteenth century, after Ferdinand Magellan's surviving crew had circumnavigated the globe. Nearing home, they were shocked to find that the day of the week they had entered in their log differed, by one day, from the day of the week on land. Like the quote from Defoe's *Robinson Crusoe* that opens this introduction and the illustration that provides the frontispiece for this book, these chroniclers began to worry that they had not kept an accurate record of the days, events, and especially the Sabbaths during their years away from "civilization."[5]

Both the explanation for the cause of the disparity in days and the need to adjust a ship's reckoning to compensate for it soon became common knowledge. Nevertheless, circumnavigations over the following two and a half centuries gave evidence that an agreed-upon place to make a change in date would reduce ambiguity in describing events and other happenings. The challenge of locating the International Date Line—an imaginary demarcation that twists and turns as it sweeps across the Pacific Ocean—came to the fore in the nineteenth century, the same era in which other aspects of time uniformity became subjects of interest.

The second part of our story, the drive to increase uniformity in time-keeping, starts in 1870, when scientists and professional societies began in earnest to address the problems created by the multiplicity of initial meridians used on cartographic products: maps and charts at all scales, sailing directions, and the predictive tables of celestial events issued annually for ocean navigators and astronomers. As we shall see, numerous proposals for a common initial meridian were advanced. Around 1879 longitude and time came together with the first systematic advocacy for a single time to be used everywhere. That common time was to be tied to the selection of one globally recognized initial meridian. A torrent of studies from different professional societies followed, offering proposals for various ways to achieve a more uniform time.

I shall also examine how the linked topics of uniform time and the selection of a prime meridian for specifying longitude were gradually taken up by central governments, and I will trace the paths by which legislation was introduced and finally passed in country after country that established a system of national times. Similarly, I will consider in detail the lengthy process by which a common meridian was adopted and comment on why the processes leading to it and to uniform time took so long and which individuals and countries caused delays.

The last part of the history of the campaigns for time uniformity introduces an entirely new concept: clock time as a social instrument. In 1907 one private individual, a London builder of fine houses, brought this idea to public attention when he proposed that clocks in England be advanced during the summer months, allowing society thereby to gain the benefits associated with having more daylight in the evenings. His proposal, which required altering public clocks twice or more each year, was opposed by many astronomers and other scientists. Yet within less than a decade, the seasonal time shifting of clocks had taken hold throughout most of Europe, in North American countries, and in the more temperate regions of the Southern Hemisphere. The process by which Summer Time became the norm for the United Kingdom and most of Europe is of particular interest today in view of the current debate in the United States and elsewhere over proposals to extend "daylight saving" time or even make it permanent.

In the United States altering public time via shifting public-clock displays has antecedents different from those articulated in Europe. The piecemeal approach to periodically advancing public clocks burdened the country with four decades of timekeeping troubles, which were not resolved until passage of the Uniform Time Act of 1966. Some of the issues encountered during those forty years of nonuniformity are addressed in the epilogue. There you will also find a brief discussion of the trade-offs, involving benefits for some and problems for others, that inevitably result when time shifts are extended for periods longer than the five summer months.

Even today complete acceptance of several uniform-time conventions has not occurred. For example, Newfoundland, India, and Afghanistan have not adopted the worldwide system of hour zones in toto, preferring instead to link their civil times more closely to the country's local time. China ignores the concept of hour zones linked to geography and maintains one single time throughout the country. And some regions in the United States and Canada do not observe annual shifts in civil time, preferring the simplicity of clocks that are not altered twice a year.

The lack of uniformity in civil timekeeping still troubles some. Inevitably, proposals to remedy this lack will be advanced. This book looks at past problems and their resolution as guideposts and warnings for future generations, who will undoubtedly struggle not only to regulate the public's time, but also to maintain cooperation when setting consistent, stable, and acceptable standards in multinational public activities.

PART I

Creating a Date Line (1522–1921)

CHAPTER I

What a Difference a Day Makes

[O]ne ought to assign a definite place where a change in the name of the day would be made.
> —Nicholas Oresme (d. 1382)

[W]e charged those who went ashore to ask what day of the week it was, and they were told . . . that it was Thursday, which was a great cause of wondering to us, since with us it was only Wednesday.
> —Antonio Pigafetta, 9 July 1522

On our arrival here [at Pitcairn Island] we found that Old Adams was mistaken in the day of the month; he considered it to be Sunday the 18th of September, whereas it was Saturday the 17th.
> —Philip Pipon, commanding HMS *Tagus*, September 1814

On Wednesday, 9 July 1522 (O.S.), the crew of the Spanish ship *Victoria* found themselves once again in the latitude of Portugal's Cape Verde Islands off the coast of Africa. The ship and its thirty exhausted and ailing sailors—survivors of Ferdinand Magellan's fleet of five ships and 265 men—had sailed westward from Seville for the Spice Islands (the Moluccas) almost three years before and were now returning home. Anchoring near the island of Santo Antão, the vessel's captain, Juan Sebastián del Cano, ordered thirteen of the deckhands to launch a small boat and when on land barter a portion of the *Victoria*'s cargo of cloves for desperately needed food and water. They were also told to find out from those living on the island the day of the week.[1]

Both Antonio Pigafetta (the diarist) and Francisco Albo (the pilot) had kept journals throughout the arduous circumnavigation. Thus they were astonished to learn that it was Thursday on the island, for their own reckoning had it Wednesday. A bewildered Pigafetta wrote, "[W]e could not see how we had made a mistake; for as I had always kept [the day and date] well, I had set [them] down every day without any interruption." Albo, however, concluded, "I believe that we had made a mistake of a day."[2]

Sometime after the ship's return to Seville, also one day later than the reckoning on board the *Victoria*, Pigafetta learned that he and Albo had made no error in their accounts. By sailing around the world westward, they had gained twenty-four hours in their reckoning, every day on board their sailing ship having been an imperceptibly small fraction longer than a so-called natural one of twenty-four hours—"as is clear to any one who reflects upon it," wrote Pigafetta later.[3]

Even though the *Victoria*'s crew were the very first to actually experience the subtle consequence of a circumnavigation, others had already explored the matter. One of the clearest and most fascinating expositions came from Nicholas Oresme (ca. 1320–1382), polymath and bishop of Liseaux, who on several occasions during the last half of the fourteenth century considered the case of two travelers, one of whom circumnavigates the earth in a westerly direction, while the other journeys eastward around the world.[4] Having returned home on the same day, Oresme observed, the first voyager would find his reckoning a day earlier than that of those who had stayed at home, while the other traveler would find his reckoning one day later. But because the same day could have three different names, he argued, "[O]ne ought to assign a definite place where a change of the name of the day would be made." Oresme's comment is perhaps the earliest reference to the need for a change-of-date line.[5]

Of course, the world's first circumnavigators experienced a result identical to that of Oresme's thought experiment. By sailing westward, Magellan's fleet continuously gained on the movement of the sun. When the fleet was 30° west of Seville, noon came two hours later than noon in Spain; when the ships were 180° west, on the opposite side of the earth, noon came twelve hours after the day's noon in Spain. Thus, when the crew of the *Victoria* again found themselves at the Cape Verde Islands, no doubt they were aware they had circumnavigated the globe—but, as we have seen, they were unaware that they had gained twenty-four hours in their reckoning.

Accounts of the first rounding of the world became available early on, notably via the historian Peter Martyr's *De Nouo Orbe*.[6] Martyr questioned the voyagers soon after their return and posited two (untenable) explanations for the discrepancy in dates. Still puzzled by their reports, which he knew had never been asserted before, Martyr conferred with the Republic of Venice's ambassador at Emperor Charles V's court, Gasparo Contarini. Trained in science and philosophy, Contarini gave the same explanation Oresme had proffered almost two centuries before. His particular example involved Spanish and Portuguese fleets leaving on the same day from the Cape Verde Islands, the former traveling westward and the latter traveling eastward. Returning to the same locale on the same day, say a Thursday on land, the Spanish would call the day Wednesday, while the eastward-sailing Portuguese would have it as Friday.[7]

Undoubtedly translations of Peter Martyr's works were scrutinized by other Europeans contemplating voyages to the two Indies; however, an English translation of his discussion of Magellan's circumnavigation did not appear until 1612.[8] So while Francis Drake himself may not have known precisely why, in 1580, he lost a day on returning from his circumnavigation—the world's second—most likely the author of the account of Drake's voyage, printed in 1628, did—a view supported by the simplicity of the text:

And the 26 of Sept. [1580 O.S.] (which was Monday in the iust and ordinary reckoning of those that had stayed at home in one place or countrie, but in our computation was the Lords day or Sonday) we safely . . . arrived at *Plimoth*, the place of our first setting forth, after we had spent 2 yeares 10 moneths and some few odde daies besides . . . in this our encompassing of this neather globe, and passing round about the world. . . .[9]

As the Spanish sailed west from New Spain, establishing their first permanent settlement in 1565 in the Philippines (which they called the Western Islands), the Portuguese continued their eastward voyages, having founded Macao in 1557.[10] In each case the colonial administrations kept the date of the home country, the Portuguese that of Lisbon, the Spanish that of Cádiz. Thus at Macao, over 122° east of Lisbon, the day's noon was a bit more than eight hours in advance of noon in Portugal. In Manila, lying over 194° west of Cádiz, the local time was more than fifteen hours behind Spain's—this despite the fact that the difference in longitude between Portuguese Macao and Spanish Manila is only 7°26', a difference in time of approximately thirty minutes.

The priests and government officials sent to administer the Philippines arrived via the once-a-year galleon from Acapulco, the route across the Pacific having been established in 1565.[11] When those from Manila met their counterparts in Macao, the former found the latter's reckoning was a day in advance of their own, as several accounts of late-sixteenth-century events document. For example, in "about 1582," a Jesuit priest, Alonso Sanches (Alfonso Sanchez) sailed from Manila to Macao and arrived there on 2 May, intending to celebrate the Feast of St. Athanasius. However, he found his Portuguese coreligionists were celebrating the Feast of the Invention [Finding] of the Holy Cross, which takes place on 3 May.[12]

The priest's story has been cited as an example of someone unaware of the difference in dates.[13] However, given that both the history of Magellan's expedition and the colonization of the Orient in different directions were so well-known by the late 1500s, it seems more likely that the Jesuit historian who first wrote about the difference was merely adding a note of realism to his discussion—an excellent one, by the way—of the reason for the discrepancy between dates.[14]

A similar sixteenth-century account is that of the Florentine merchant Francesco Carletti, who sailed with his father from San Lucar, the same Spanish port from which Columbus and Magellan embarked on their voyages of discovery. Arriving in New Spain, the Carlettis crossed the Isthmus of Panama to the Pacific Coast and on 25 March 1596 (N.S.) sailed from Acapulco on that year's Manila Galleon. They arrived at Cavite in the Philippines in June and in May of the following year took passage for Nagasaki, Japan, where the Portuguese from Macao were struggling to maintain a foothold. There the Carlettis encountered the difference in dates. The merchant's explanation, both correct and matter-of-fact, reinforces the judgment that this discrepancy was understood by educated Europeans generally, and certainly by navigators charged with keeping their ship's daily log.[15]

Later voyagers and travelers also noted the difference in days. In October 1616, Jacob Le Maire and Willem Schouten arrived at Batavia, on the island of Java. Sailing westward from Holland, they had been the first to find open passage around the tip of South America, a discovery that the representatives of the Dutch East India Company did not believe. Le Maire remarks on the conflicting dates in reporting the seizure of his ship by the officials, who had come eastward from the Netherlands.[16]

One more account of the difference in days between Spanish and Portuguese possessions in the Far East—and one that provides an experimentally verifiable explanation—is that of Giovanni Francesco Gemelli Careri, an Italian doctor of civil law. Traveling eastward to Asia from Europe, Gemelli Careri eventually sailed from Macao to Manila. On his arrival there, he asked and was told the day was Monday, 7 May 1696. His own running account, though, gave it as Tuesday, 8 May.

Professing to be amazed by the difference in dates, but more likely wanting to demonstrate his complete understanding of the phenomenon, the learned traveler called attention to the different routes routinely taken by the Portuguese and Spanish mariners. Then, referring to tabulations of the sun's daily declination, which are based on particular meridians, he posited two sailing ships leaving from the meridian of Lisbon, carrying with them identical declination tables. If two ships, one sailing eastward and the other westward, left Lisbon on 1 May 1630 and both returned exactly one year later, that day (1 May 1631) would be a Thursday for those who remained at the port. The table gives the sun's noon declination as 15°6', a value Gemelli Careri confirms as "true and immutable."

But, he notes, although the daily account of the westward-sailing navigator will be one day less, the sun's declination for Wednesday, 30 April 1631, printed in his table is 14°48'—which is *not* the value he would actually observe on his return to Lisbon. Similarly, while the eastward sailor's shipboard account would show the date of Friday, 2 May 1631, on his return, his table would give the sun's declination for that day as 15°24'—a measurable increase of 18' over what he would actually observe. After giving this explanation, Gemelli Careri adjusted the reckoning in his own account by simply repeating Monday, 7 May 1696.[17]

The significance of having the correct day throughout a voyage was underscored by William Dampier in *A New Voyage Round the World*. Having sailed west across the Pacific keeping the Western dating, and now (in January 1687) departing from the Philippine island of Mindanao, this navigator-adventurer reports that the day on the island was that observed everywhere in the East Indies—and the same used by Europeans sailing to the Orient via the eastward route. After providing the usual explanation of leading and lagging the sun, Dampier writes:

One great Reason why Seamen ought to keep the difference of time as exact as they can, is, that they may be more exact in their Latitudes. For our Tables of the Sun's declination, being calculated for the Meridians of the places in which they were made, differ about 12 Minutes from those parts of the World that lie on their opposite Meridians, in the Months of *March* and *September*. . . . And should they run farther as we did, the difference would still increase upon them, and be an occasion of great Errours.[18]

Since latitude is determined via a noontime observation of the sun's altitude above the horizon, any error in finding the tabulated declination on that day directly affects the result, leading to a latitude error that can reach nearly twenty miles. In this era, which lacked any means of determining longitude at sea, latitude sailing was the norm: keeping to a latitude while journeying east or west and then sailing south or north when one's estimate of longitude "by account" had been reached.

It is of course unnecessary to change the date used in a ship's log when the "other" date is encountered at some distant place. All one needs is consistency and a proper understanding of the sun's declination. Still, it was easy to make errors. Dampier noted this, writing that

Seamen in these Voyages are hardly made sensible of this, tho' so necessary to be observed . . . as it happened among those of our crew; who after we had past of 180 Degrees, began to decrease the difference of declination, whereas they ought to have increased it, for it all the way increased upon us.

The logs and associated journals of many voyages of the eighteenth century show that it was common practice to change the date before a ship returned to its home port. Among English circumnavigations were those captained by Woodes Rogers in 1710 (who added a day at Batavia), George Anson in 1742 (who added a day at Macao), James Cook in 1770 (who added a day at Batavia), Cook again in 1775 (repeating a day at Cape Town), and George Vancouver in 1795 (who repeated a day at St. Helena) (see Table 1.1).[19] This last explorer-hydrographer explained:

[B]ut as it was now become expedient that we should subscribe to the estimation of time, as understood by Europeans and the rest of the civilized world, to which we were now fast approaching, our former reckoning was abandoned, the day we had gained was dropped, and, after noon this day, we recommenced Sunday the fifth of july.[20]

Had the Philippines remained the only European colony on the "wrong side" of the 180th meridian—its status for two hundred years—the anomaly could simply have been dismissed with a footnote in a compendium of sailing directions. But British and Russian explorations in the

TABLE 1.1

Finding a difference in dates during early exploration and travel

Who	Where	When
Juan Sebastián del Cano	Cape Verde Islands	1522
Francis Drake	[Home]	1580
Alonso Sanches, S.J.	Macao	ca. 1582
Thomas Cavendish	?	1588
Francesco Carletti	Nagasaki	1597
Jacob Le Maire	Batavia (Djarkarta)	1616
William Dampier	Mindanao	1686
Giovanni Gemelli Careri	Manila	1696
Woodes Rogers	Batavia	1710
George Anson	Macao	1742
Louis Antoine de Bougainville	Île de France (Mauritius)	1768
James Cook	Batavia (Djarkarta)	1770
James Cook	Cape Town	1775
George Vancouver	St. Helena	1795

eighteenth century necessitated some important changes. In the Pacific, for example, James Cook's explorations increased beyond measure European fascination with the region. Exotic islands, friendly and unfriendly natives, and hitherto unknown flora and fauna all waited to be exploited and/or taught Western European ways of behavior and religion.

In October 1788 Lieutenant William Bligh, captain of HMS *Bounty*, sailed into Tahiti's Matavai Bay. This was his second journey to the island: he had been there in 1777 with Cook during that distinguished explorer's ill-fated third voyage. The *Bounty* and its crew had been sent to Tahiti by the Admiralty to collect breadfruit plants for subsequent transport to British islands in the West Indies, where the plantation owners hoped the mature trees' fruit would provide a nutritious staple for the slaves working their fields.

The *Bounty*'s voyage had been extremely arduous. Sailing from England late in December 1787, the ship did not reach Staten Island in the South Atlantic until the end of March, far too late for a comfortable passage around the tip of South America. As the crew sailed farther south, they encountered the terrible storms for which the Cape Horn region is famous. After struggling for thirty days to round the Horn and sail into the Pacific, the *Bounty* was starting to leak more and more. So Bligh gave orders to turn away and sail to the Cape of Good Hope for repairs and reprovisioning. Thus, when the ship finally reached Tahiti six months later, it had sailed to the South Pacific island not westward as planned, but via the eastern route; its resulting longitude was 210°30' east of Greenwich, or slightly more than fourteen hours ahead of the time at home.[21] Throughout their nearly six months on shore at Tahiti, the *Bounty* crew continued to keep the Eastern or Asian dating they had brought with them.

After loading the breadfruit trees, the *Bounty* left Tahiti sailing west, bound for the plantations in the British West Indies. Following the mutiny of 28 April 1789, the vessel returned to Tahiti with the mutineers and those who could not be placed in the *Bounty*'s overloaded launch with Bligh. Subsequently, one of the latter wrote of his twenty-one months ashore on Tahiti, "We kept the Hollidays in the best manner that we could, killing a Hog for Christmas dinner, and reading prayers which we never omitted on Sundays."[22]

The *Bounty* sailed from Tahiti on 23 September 1789 with nine mutineers and their Tahitian wives; six male Tahitians brought as servants, three of whom also had wives; and one child. For over eighteen years no one knew their whereabouts. Then Mayhew Folger, master of the sealer *Topaz* out of Boston, happened by Pitcairn Island on 6 February 1808. Years of turmoil and murders among those who had fled to the island had produced one of the most fascinating and unusual communities in the world. Folger reported the island population consisted of seven widows and more than two dozen children, all led by the one surviving mutineer, John Adams (also known as Alexander Smith).

Like Robinson Crusoe, the islanders had lost their reckoning during their years of isolation. Folger wrote in his log that the mutineers "ran the ship Bounty on shore and broke her up which took place as near as he [Adams] could recollect in 1790."[23] Most certainly the Pitcairners regained their daily reckoning from Folger, who had sailed the eastern route into the South Pacific (see Figure 1.1). As a result, those on Pitcairn began keeping the Asiatic or Eastern date, even though the island was 130°00' west of London by Folger's lunar observations—and thus fifteen hours and twenty minutes in advance of Greenwich time.[24]

That the *Topaz* had sailed via the eastern route was not known to the next visitors to Pitcairn Island, Sir Thomas Staines and Philip Pipon, respectively commanders of HMS *Briton* and HMS *Tagus*. Sailing from Valparaiso near the end of the War of 1812 and cruising the Pacific to protect British whalers, early on Saturday, 17 September 1814, lookouts on the frigates unexpectedly spotted Pitcairn, which had been incorrectly located on charts ever since its discovery in 1767.

Astonished to find people living there, the two commanders went ashore late that morning. There they found the surviving mutineer John

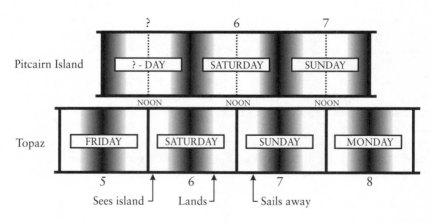

FIGURE 1.1

In early February 1808 Mayhew Folger, master of the American sealer Topaz, *came upon the last of the HMS* Bounty *mutineers and the women and descendants of those mutineers who had fled to Pitcairn Island. The community had lost its daily reckoning and recovered it from Folger.*

Adams and the rest of a deeply religious community.[25] They also found a difference in dates. Captain Pipon remarked:

On our arrival here we found that Old Adams was mistaken in the day of the month; He considered it to be Sunday the 18th of September, whereas it was Saturday the 17th: and to his credit they were keeping the Sabbath very properly & making it a day of rest & prayer. By his account he had been misled by the Captain of the Topaz.[26]

While blaming the American Folger may only have been a ploy by Adams to curry favor with the British naval officers (who held his life in their hands), his words reinforce the conclusion that the islanders lost their daily reckoning sometime after 1790 and did not recover it until 1808. Then in 1814, the Pitcairners shifted their reckoning to the Western or American dating (see Figure 1.2).[27]

Contact with the isolated Pitcairn community marked a new chapter in the history of confusion regarding the shifting of days at the boundary between the West and East.[28] On board HMS *Briton* Lieutenant John Shillibeer, Royal Marines, had been unable to go ashore. Nevertheless, his secondhand statement that John Adams had been keeping a daily reckoning one day ahead of the ship's log is consistent with Captain Pipon's report.[29] Unaccountably, however, in referring to the mutiny leader Fletcher Christian's son, Thursday October Christian, who came on board the *Briton* and was sketched by Lt. Shillibeer, the officer gives

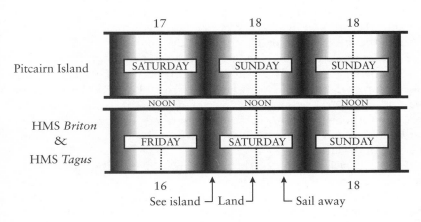

FIGURE 1.2

In September 1814, near the end of the War of 1812, two British warships sailing out of the South American port of Valparaiso chanced on Pitcairn Island. The ship captains found that the community living there was one day in advance of their own reckoning. As shown in the figure, the Pitcairners switched from their Asiatic reckoning to the American one by repeating a day.

his name as "*Friday* October Christian." If this adjustment was done to make the name of Pitcairn's first-born conform to the Western or American date, the sketch should have been captioned "*Wednesday* October Christian." The name change in Shillibeer's account (which gained wide circulation) was to bedevil a host of subsequent writers.[30]

Shillibeer's confusion of days was not the only one resulting from this 1814 encounter. After the various reports regarding the British discovery at Pitcairn Island had been received at the Admiralty, the permanent secretary, John Barrow, summarized events in the *Quarterly Review*.[31] After noting the differing dates, he explained:

This [difference] was occasioned by the Bounty having proceeded thither by the eastern route, and our frigates having gone to the westward; and the Topaz found them right according to his own reckoning, she having also approached the island from the eastward. Every ship from Europe proceeding to Pitcairn's island round the Cape of Good Hope will find them a day later—as those who approach them round Cape Horn, a day in advance, as was the case with Captain Folger and the Captains, Sir T. Staines and Pipon.[32]

Barrow was unaware that the *Topaz* had sailed around the Cape of Good Hope following the eastern route and so had approached the island *from the west*. Had he been aware of the *Topaz*'s route, he might have given weight to Captain Pipon's observation that the date of the *Bounty*'s arrival at Pitcairn was unknown—a sure indication that the islanders had lost their reckoning sometime between 1790 and 1808 and regained it from Captain Folger. Further, Barrow assumed that the islanders did not shift their reckoning to the Western dating when the British officers brought the difference in dates to their attention.

Even assuming that the island's dating was not changed in 1814, Barrow's statement that "[e]very ship from Europe . . . round the Cape of Good Hope will find them a day later" is wrong; Pitcairners would have the *same* day as those on ships approaching from the west.[33] John Barrow's confusion—to add or subtract a day when crossing the boundary between West and East—continues to bewilder those who first encounter this line of demarcation.

The 1814 visit by the British frigates transformed the dating on an isolated Pacific island that had been observing the Eastern or Asian dating. Even so, this mode of reckoning the days had gained acceptance in the South Pacific, and over the next decades its adoption expanded. The beginnings of this expansion can be traced back to Sunday, 5 March 1797, when thirty missionaries of the newly formed Missionary Society arrived at Tahiti on the *Duff*, a ship owned by the London-headquartered society. Having departed from England the previous September, the ship arrived at the tip of South America in late November, just prior to the prudent season for doubling Cape Horn. After two weeks of severe weather, the *Duff* turned away from its attempt to enter the Pacific Ocean via the western route. And so, like HMS *Bounty* a decade earlier, the missionaries came to the South Seas via the eastern route.[34]

The proselytizers' first years on Tahiti were extremely discouraging ones; not until 1815 did they begin converting Tahitians to Protestant Christianity. Their eventual success led to missions elsewhere within the region, and several more Protestant religious organizations sent their own missions to other South Pacific island groups.[35] In light of this activity, by the 1830s it was becoming helpful to sketch a line to delineate the boundary between the Eastern (Asiatic or European) and the Western (American) date, rather than drawing circles around scattered islands.

Obviously a boundary of sorts already existed elsewhere in the Pacific. Following Cook's discovery of the Hawaiian Islands in the late 1770s, British and American whalers expanded their operations in the Pacific Ocean, bringing the Western date to those islands where they stopped to repair their ships and reprovision. As indicated in Figure 1.3, many, perhaps even most, of those who anchored in the Marquesas Islands were already observing the American dating before the nineteenth century began.

Further extension of Eastern dating in the Pacific ended in the 1830s. In the third quarter of 1834, three missionaries sent by the Missionary Society arrived at the island of Thauata in the Marquesas by way of Tahiti. It was the society's fourth attempt since 1797 to bring the Protestant Christian religion to these natives, and this latest effort would fare no better than its predecessors. Their attempt to introduce the Eastern dating also failed, as they reported the following August.

On arriving the missionaries had established the same dating used on Tahiti. However, the first ships that touched at Thauata to reprovision used the American (Western) dating, the reckoning on the Sandwich Islands and also the one used throughout the Marquesas. The conflict between the two reckonings was obvious; one captain of a British whaling ship even stated that he would observe no Sabbath but his own. On one of his Saturdays he sent in casks to be filled with water, employing natives to assist the sailors, thus making it impossible for the missionaries to hold religious services. Soon after the first ships had departed, the LMS missionaries switched to the American dating.[36]

The arrival of French Catholic missionaries led to further shifts in the boundary between the two modes of dating. Sailing into the South Pacific via the westward route, in 1834 they established a mission at Mangareva in the Gambier Islands, and in 1836 they attempted to establish a

Creating a Date Line

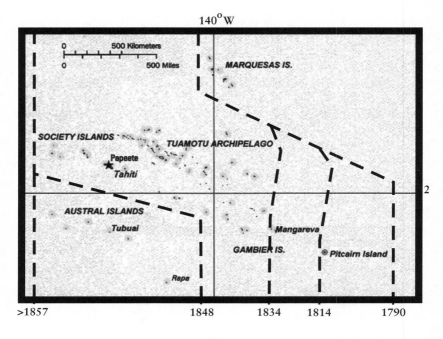

FIGURE I.3

During the 1800s the boundary between the Asiatic and the American dating moved to the west, starting from Pitcairn Island in the South Pacific at 130°W longitude. It continued moving west, albeit slowly; Stieler's Hand-Atlas shows it west of the Society Islands in 1892, at the dashed line at the left of the figure.

mission in Tahiti. They were repulsed by the English missionaries and the native government, which the missionaries controlled. The French government responded in 1838 by sending a warship and issuing demands for compensation. A French protectorate was proclaimed a few years later, and in 1847 the island's protectorate status was at last recognized by Tahiti's Queen Pomare, who had been battling the annexation. In 1848 the American dating was introduced by the French, and, as a later historian observed, "Tahiti ceased to be a day in advance of the rest of the world."[37]

The isolated Austral Islands eventually became part of French Polynesia, and its Eastern dating was altered to conform with the rest of the French-dominated area of the South Pacific. Exactly when this shift occurred is not known; however, in September 1857 two LMS missionaries, Alexander Chisholm and Charles Barff, left Raiatea, one of the Society Islands, for Tubuai in the Austral Islands, arriving there on Saturday evening, the twenty-sixth of the month (by the American dating). In his report, Chisholm noted that when they went ashore the next day, which was their Sabbath, they found it "to be their [Tubuai's] Monday morning."[38]

While France was consolidating its authority and instituting the American dating over much of the region, a shift in quite the opposite direction took place in the far western Pacific. As noted earlier, during the 1600s the use of the American date in the Philippine Islands did not extend throughout the archipelago, for the Muslim-dominated areas in the south observed the Eastern date. Over the next two hundred years the colonial authorities in Manila exerted their influence over more and more of the islands, bringing to the people there Catholicism and the government's official date. But even as late as 1851, when the central government finally captured the town of Jolo on the Sulu Sea island of that name, large regions of the island of Mindanao and elsewhere were only nominally under their control. (In spite of the historical record, maps show the entire archipelago observing the American date.)

The decree by the governor-general of the Philippines that instituted the change was issued on 16 August 1844 (American date) and reads:

The Government of the Philippines. Inasmuch as it is deemed useful that the way of reckoning dates in these Islands be consistent with the system used in Europe, China and the other countries situated east of the Cape of Good Hope, which reckon the day ahead for reasons well known to us, I direct after agreement with His grace the Archbishop, that this year only, the Tuesday, December 31 be dropped, as if it had really passed and that day following Monday December 30 be reckoned as Wednesday January 1, 1845, which will be the beginning of the calendar of that year in which no alteration is to be made.[39]

Despite the announcement and the resulting alteration, many countries' encyclopedias and geography books continued to state that civil times in Macao and Manila differed by twenty-four hours. A particularly egregious example is American publications for schoolteachers, the accompanying maps of which show the Philippine Islands on the Western (American) side of the International Date Line, a faulty representation that continued into 1891.[40] As we shall see, some repercussions associated with this error came during the late 1890s.

The issue of changing one's daily reckoning when sailing on the Pacific Ocean is different from that of adopting a particular dating on land. As we have seen, William Dampier—and undoubtedly others as well—early on urged changing dates at the same time a navigator changed his entries from west to east longitudes, and vice versa. With the annual issuance of the Greenwich-based *Nautical Almanac*, first published in 1766 for the year 1767, and the concurrent development of the finding of longitude from lunar observations and/or via the marine chronometer, ocean navigation altered. One of the era's first changes, particularly on naval vessels, was to recast the ship's daily log by switching its entries to the civil day, which begins at midnight, thus making obsolete use of the nautical day, which begins at noon of the previous day.[41] Changing the day itself on crossing the anti-meridian of Greenwich eventually became common practice as well. However, no official directives have been located, so one must rely on the actual ships' logs to determine approximately when this change took place.

Perhaps the most famous transpacific passages by English-speaking sailors were the American trading voyages of the early 1800s. Most of them were circumnavigations: around Cape Horn to the Pacific Northwest to barter for furs; then to Canton via Hawaii, where the furs were traded for Chinese tea, silks, and porcelains; and finally home via the Indian and Atlantic Oceans.[42] A review of the surviving logs of these merchant vessels over the period 1800–1830 shows great variability in captains' decisions regarding the change of date. Out of a total of eleven voyages, three ships dropped a day at Canton to conform with the Eastern dating there, three dropped a day when passing the 180th meridian from Greenwich, one dropped a day upon returning home, and four ship logs that ended at Canton continued to keep their positions in terms of west longitude after passing the 180th meridian.[43]

Nor did U.S. Navy ships maintain any consistency at first. In December 1829 the USS *Vincennes*, which circumnavigated via the western route, dropped a day as it approached Macao; and in October 1832 the U.S. frigate *Potomac*, which also circumnavigated the globe, repeated a day at Valparaiso, Chile, after crossing the Pacific via the eastern route. By the late 1830s, however, consistency among American naval vessels was the norm. The U.S. frigate *Columbia*, for example, repeated a day when it crossed the 180th meridian on 26 September 1839, and Charles Wilkes, commander of the 1838–1842 U.S. Exploring Expedition, altered his date a total of three times on his multiple crossings of the 180th meridian.

Reinforcing the view that uniformity in approach among navigators using the Greenwich-based *Nautical Almanac* probably came in the 1840s is the difference in practice between Robert Fitzroy, commander of HMS *Beagle*, and James Ross, who explored the Antarctic regions. In November 1835 the former, sailing the westward route, changed his reckoning from 16 to 17 November 1835 at Tahiti "to agree with the reckoning . . . of those [the LMS missionaries] who came from the west"; the latter, who sailed the eastward route, had two Thursdays, 25 November 1841, on crossing the 180th meridian, "in order that our time might correspond with that of England."[44]

Returning to land, in the northern regions of the Pacific Basin one encounters a reckoning different from the rest of the Western European–dominated islands. Early Russian explorations in Siberia and later across the North Pacific Ocean led, in 1741, to the discovery of Alaska and many of the Aleutian Islands. Exploitation of the region's resources began with the hunting of seals, sea otters, foxes, and other fur-bearing mammals. By the end of the eighteenth century, the abuse and even killing of the Indians by Russian hunters in their quest for furs had begun to concern the czar's government. Partly as a way to control these excesses, the Russian-American Company was created as a quasi-government corporation with monopoly rights throughout the region. When the government failed in its attempts to make the North Pacific off-limits to all but Russians by fiat, formal conventions negotiated with the United States (1824) and Great Britain (1825) fixed the boundaries of Russia's North American possessions to a narrow strip of the northwest coast northward from 54°40', the Alaskan peninsula, and the islands west of the 141st meridian from Greenwich.[45]

With the Russian-American Company had come priests of the Russian church and their calendar, which was still based on the Julian dating. By 1800 the difference between the Julian and Gregorian reckoning amounted to twelve days, so the 1825 Convention between the King of the United Kingdom of Great Britain and Ireland and the Emperor of all the Russias carried two dates—28 February 1825 (N.S.) and 16 February 1825 (O.S.)—as had the slightly earlier one between the United States and Russia.

Because Russia had expanded eastward from the Asian continent, Russian Alaska used the Eastern dating. As in all other places in the world at the boundary between the two systems of dating, problems occurred. For example, during his 1837 layover at New Archangel (Sitka), the headquarters of the Russian-American Company, Commander (later Admiral) Edward Belcher, commanding HMS *Sulphur*, wrote:

The [Russian] colony having arrived from the westward, brought their own Sunday; consequently we were generally working on our opposite holidays, a measure I could only obviate by respecting their day of worship, and giving our men a holiday. To our artificers, who could not work at the dockyard on their Sabbath, this was a serious drawback, when we considered the short period of our stay.[46]

Throughout this period the Russian-American Company's hunters, in their search for seals and especially for the extremely valuable sea otters, ranged farther and farther south, reaching the San Francisco Bay area early in the nineteenth century. With only one ship a year coming from St. Petersburg, it was proving extremely difficult to assure sufficient provisions for those in Alaska. The company therefore decided to establish an outpost on the California coast in order to produce food for those living in the far north. Two sites north of San Francisco were acquired, one at Bodega Bay, eighteen miles west of present-day Santa Rosa. On 15 March 1812 (N.S.), the construction of what would become Fort Ross got underway at a second coastal site, located about eighteen miles north of the first one.[47] Fort Ross was dedicated on 11 September 1812 (see Figure 1.4). Its settlers numbered about one hundred Russian and eighty Aleutian hunters. At first sea otters were hunted, a most profitable enterprise, but within a few short years the local population had been hunted almost to extinction. The colony then turned in earnest to farming, its primary purpose. However the Russians' agricultural efforts were a near-failure and produced only meager supplies for the several thousand Russians and supporting workers in Alaska.

One of the visitors to Fort Ross was Ioann Veniaminov, a priest. In 1836 he gave its population as 260 persons—154 males and 106 females—of whom 120 were Russians, the others being Creoles, Aleuts, and baptized Indians. Journeying to the San Francisco Bay area as he waited for passage back to Sitka, he stayed with the priests and monks at Mission San José. There he noted the difference between his Monday and "Sunday by their calendar."[48]

Seeing no advantage to continuing the Fort Ross enclave, the Russians finally abandoned it. The buildings and assets were sold to John Sutter of New Helvetica and the property deeded to him. In December 1841 the Russians and their associates left the site and sailed to Sitka, taking their language, their religion, their Julian calendar, and their Eastern dating with them.[49]

The next shift in the boundary between east and west came twenty-five years later with the purchase of Russian Alaska by the United States. The Convention for the Cession of Alaska was signed in Washington in March 1867, ratified first by the United States and then by Russia, and proclaimed on 20 June 1867.[50] The formal transfer of territory took place on 18 October 1867 (N.S., American date), as reported in San Francisco's *Daily Alta California* (in a number of articles published between 26 October and 20 November 1867) and elsewhere.[51] With the transfer of sovereignty the Eastern dating shifted to the Western one in all the former Russian territories, including the Aleutian Islands west of the 180th meridian.

However, unlike in the Philippines, where adjusting the ecclesiastical calendar took place simultaneously with the change in the civil date, in Alaska two Sabbaths continued to be observed every week. According to George Davidson, a contemporary field geodesist and astronomer who

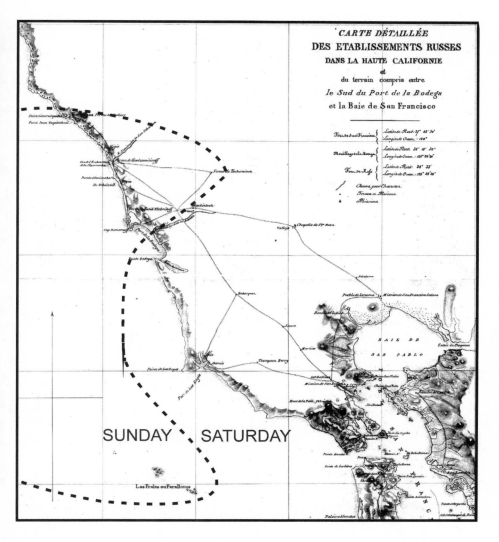

FIGURE 1.4

*The Fort Ross settlement included nearby ranches, a deep-water anchorage at
Bodega Bay, and a permanent hunting camp on the Farallon Islands. The dashed
curve shown on the manuscript-map segment gives the approximate boundary
between the Russians' Asiatic dating and the American one. (The Russians also
used the Julian calendar.) The map is in the Duflot de Mofras collection, courtesy
of the Bancroft Library, University of California, Berkeley.*

investigated the matter, "[T]he Holy Orthodox Church in Alaska did not change until January 1st 1871 by order of Right Reverend John [Ioann Mitropol'skii], Bishop of Alaska & the Aleutian Islands."[52] A long chapter in Alaskan history was over.

Without question, the 1872 appearance of Jules Verne's *Around the World in Eighty Days* sensitized Westerners to the date-change consequences of circumnavigating the globe.[53] Phileas Fogg's eastward journey and the just-in-time revelation that he had gained a day still charms and even surprises some readers, in an age when going around the Earth can easily be accomplished in a very few days.

The issue in the real world was not gaining or losing a day to win a wager. It was the need to articulate which dating, Eastern or Western, people would live by. For example, the Fiji Islands straddle the 180th meridian from Greenwich, now the common location throughout the Pacific for altering the date. By custom, both the European and native populations on both sides of this meridian reckoned in terms of the Eastern dating. Thus on 5 June 1879 the governor of the recently established British crown colony (1874), upon the advice and consent of the legislative council, enacted an ordinance "to provide for a universal day throughout the colony," thereby legalizing traditional practice.[54]

A dozen years later yet another shift of the date line took place. At that moment in its history the islands of Samoa were being governed under a tripartite protectorate created by imperial Germany, Great Britain, and the United States. The islands, though east of the 180th meridian, used the Eastern dating, which had been brought from Australia. Steamships from Australia and New Zealand and also from the United States regularly stopped at the Samoan port of Apia. An American businessman urged Samoa's nearly powerless king, Maileota Laupepa, to change to the Western dating, giving commerce as his reason. The shift would give Apia the same day and date as the steamers coming from San Francisco. (A secondary reason, which pleased the European religious interests on Samoa, was that these scheduled ships would no longer arrive on a Sunday, thus removing the need for Samoans to work on the Sabbath.) Although the Germans opposed the shift, King Maileota ordered that two Mondays, both 4 July 1892, be observed, thus bringing Samoa's dating into line with the Western (American) system of dates.

Some accounts of this shift suggest that the choice of 4 July was a diplomatic maneuver designed to flatter the Americans; however, no evidence of this has been found.[55] Moreover, it is not clear how soon the commercial activities in Samoa actually adopted the change. One account written six months later mentions a steamship at anchor at Apia that was still using the Eastern date.[56]

Throughout the last quarter of the nineteenth century, articles frequently appeared purporting to depict the sweep across the Pacific of what was being called the date line. Several of them turned out to be incorrect. Still others "got it wrong" when describing the change of day crossing the 180th meridian. Furthermore, the changes in dating practices on various islands meant that geographic atlases were soon out of date, some badly so.[57] What became the first authoritative determination of the date line's sweep occurred soon after Commodore (later Admiral) George Dewey's naval victory at Manila Bay—and the battle reports telegraphed to the United States.

In mid-May 1898, two weeks after the American victory, a California congressman received a letter asking for authoritative information as to Manila's time reckoning. The writer pointed out that the 1887 edition of *Harper's School Geography* depicted the International Date Line with the Philippine Islands lying to its east. The writer was sure this could no longer be the case and asked when the shift in date had taken place, for the reports stated that the American naval engagement with the Spanish fleet took place on Sunday morning, Manila time (which was Saturday afternoon in the United States).

This letter was forwarded to the U.S. Navy's Hydrographic Office. Its official response was to send a four-year-old, two-page "Memoranda [*sic*] on the International Date Line." It began: "It appears that no information concerning the actual date in local use on each of the Pacific Islands has as yet been collected and published in such authoritative form as to be entitled to entire confidence," with the further remark that "the international date line has not . . . been a matter of record in the Hydrographic Office, nor do the sailing directions published by this and by European nations take cognizance of it."[58] Given this less-than-excellent response to a member of Congress, a proposal to the hydrographer the following year must have seemed a godsend.

In a letter dated 3 March 1899, George Davidson of San Francisco (1825–1911) offered to draw the date line through the Pacific on a copy of the Hydrographic Office's submarine cable chart, "if you care to receive such information," stating that he had taken "some pains to learn the [change-of-date] practice of our Commercial Steamers & those of France, Germany & England."

Davidson's credentials were impeccable. A highly regarded scientist, he had joined the U.S. Coast Survey in 1845, had come to California in 1850, and had eventually become assistant-in-charge of the hydrographic and geodetic surveys along the West Coast and in Alaska. Dismissed from the Coast and Geodetic Survey without cause in 1895 during the second Grover Cleveland administration's purge of survey staff, Davidson became a professor of geography at the University of California, continuing his longtime ties with that institution (see Figure 1.5). He served for many years as president of the California Academy of Sciences and was also a member of the National Academy of Sciences in Washington.

The hydrographer Captain Joseph E. Craig accepted Davidson's offer at once and had a copy of the recently issued official chart sent to him. On it the geographer drew in blue the sweep of the date line and returned the chart in early April.[59] (The line is dated June 1898, most likely indicating the completion date of Davidson's investigations.)[60] The Hydrographic Office then began the preparation of a chartlet, "The Line Separating the Lands of the Pacific Where American Date Is Kept From Those Where Asiatic Date Is Kept," printing it on the "August [1899] Pilot Chart of the North Pacific Ocean" (see Figure 1.6).[61] On this first official depiction of the date line, Davidson's contribution was acknowledged by placing "According to the Investigations of Professor George Davidson of the University of California" after the chartlet's title.

While the Hydrographic Office was preparing this chartlet, the *New York Tribune* decided, no doubt independently, to feature an article on the date line. In what can only be termed a blunder, the newspaper approached William Harkness of the U.S. Naval Observatory for information. Nearing mandatory retirement age, Harkness was serving as the observatory's astronomical director as well as director of the Nautical Almanac Office during a period when one of the not-infrequent civilian-versus-military control battles was in full swing, a conflict that required his full attention.[62] Unaccountably, instead of forwarding the *Tribune*'s

FIGURE 1.5

The American surveyor and geodesist George Davidson, at an 1894 USC&GS conference on geodesy. NOAA Central Library, theb3631.

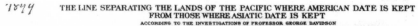

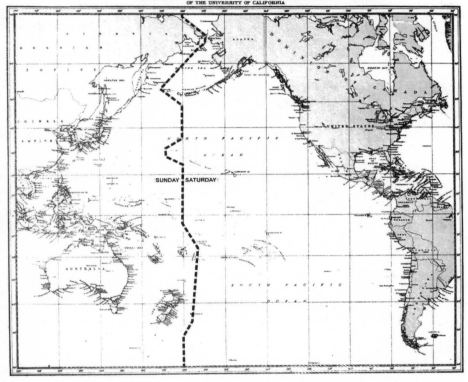

1879

THE LINE SEPARATING THE LANDS OF THE PACIFIC WHERE AMERICAN DATE IS KEPT
FROM THOSE WHERE ASIATIC DATE IS KEPT
ACCORDING TO THE INVESTIGATIONS OF PROFESSOR GEORGE DAVIDSON
OF THE UNIVERSITY OF CALIFORNIA

SUNDAY SATURDAY

FIGURE 1.6

*The U.S. Hydrographic Office chartlet based on George Davidson's investigations.
The line showing the boundary between the Asiatic and the American dating has
been enhanced. From H.O. No. 1401, Pilot Chart of the North Pacific Ocean,
August 1899.*

request to the hydrographer, Harkness answered it himself, thereby compounding the newspaper's initial blunder.

Early in May the *Tribune* published "Where the Day Begins," crediting Harkness with the map accompanying the lengthy text. Unfortunately, the map was wrong. Apparently Harkness had simply used an obsolete map, out of date by seven years, as the basis for his response.[63]

The first authority to spot this misleading article is not known. In any event, after conferring with the navy admiral directly superior to the hydrographer and receiving a letter from Davidson detailing his investigations, the *Tribune* published "Where the Day Changes." Stressing that this article, which summarized "Naval, Commercial and Local Usage in the Pacific," was based on an official U.S. Navy chart, the newspaper included a map comparing the date line provided by William Harkness three months before with the one determined by George Davidson. The differences between the two were striking. Moreover, it was perfectly clear which line was to be used. Thus, without too much embarrassment to the parties concerned, a "final verdict on the subject"—to use the *Tribune*'s words—was rendered.

Ironically, George Davidson's investigation had not been exhaustive, and so the Hydrographic Office's chartlet was also wrong when it was issued at the end of July. For one thing, the legislature on Rarotonga in the Cook Islands did not vote to shift to the Western dating until 6 August 1899. Moreover, the people on Penrhyn, 734 miles to the northeast, ignored the agreed-upon shift, a position that required them to observe two Christmases in 1899 in order to put things right.[64]

However, by 1900 the issue of who was right and who was wrong was no longer of any import. By then the British and American Hydrographic Offices were in essential agreement regarding the date line, as indicated by a map published by the superintendent of the (British) Nautical Almanac Office.[65] Two minor differences existed, however, both in the North Pacific: the American line depicted two false islands (Morell and Byers) northwest of the Hawaiian chain, and the issue of the date on Wrangell Island in the Arctic Ocean was not resolved. The first was resolved by 1911 when the false islands were removed from charts; the latter was settled in the 1920s when the Soviet Union asserted its sovereignty.[66]

In 1921 the Admiralty's Hydrographic Department prepared "Notes on the History of the Date or Calendar Line," a summary of its earlier actions that was later published in a New Zealand journal.[67] The accompanying map showed the sweep of the date line at five particular times over an eighty-year period. Copies of the map received wide circulation, unfortunately without taking into account the warning placed on it: that sections of the date lines, especially those shown near and south of the equator, were "Approximate Only."[68] The summary charts given in this chapter, with the supporting citations, provide a more complete depiction of that history (see Figure 1.7).

Furthermore, it is critical to remember the words written by George Davidson in 1901 when he introduced the linked subject of "gaining a day" and "losing a day":

There is no international date line. There is a theoretical date line through the Pacific ocean dependent on the adoption of a prime meridian through Greenwich or some place in Western Europe. (*San Francisco Examiner*, 8 June 1901)[69]

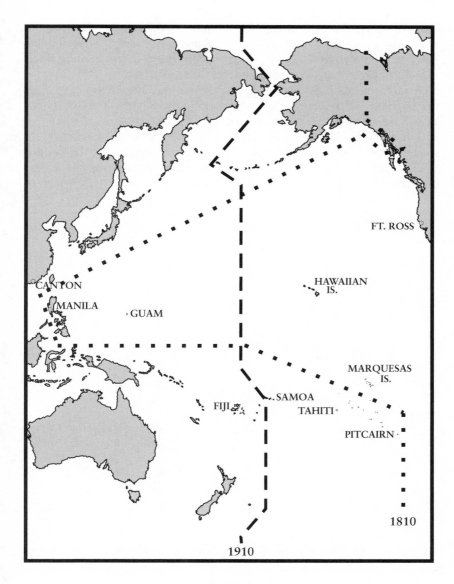

FT. ROSS

HAWAIIAN
IS.

CANTON

MANILA

GUAM

MARQUESAS
IS.

FIJI   SAMOA
          TAHITI

PITCAIRN

1810

1910

FIGURE 1.7

*The boundary between those lands using the Eastern (Asiatic) date and those
employing the Western (American) date. The 1810 line is based on the research
described here. The 1910 line reflects adjustments made by the Admiralty's
Hydrographic Department in that year, as reported in* New Zealand Journal of
Science and Technology *11 (April 1930): 385–388.*

The Admiralty belabored Davidson's main point in 1921, writing, "It should be noted that the term 'date-line' does not imply that such a line has ever been definitely laid down, either by any one Power or by international agreement."[70] The same holds true as of 2006. On a map or chart, the International Date Line is merely a conventional line depicting the boundary between those nations using the Eastern (Asian) dating and those using the Western (American) dating.

In addition, Davidson's words remind us that an agreement on an initial or prime meridian did not exist in the first years of the twentieth century. The history of that fundamental subject begins in the next chapter.

PART II

Campaigning for Uniform Time (1870–1925)

CHAPTER 2

Choosing an Initial Meridian

It is desirable that all the nations of Europe, in place of arranging geographical longitude from their own observatories, should agree to compute it from the same meridian.

—Pierre Laplace, ca. 1800

All these differing initial points make geographical charts less convenient to consult; they compel unceasing calculations, on account of substituting one number for another in the estimation of distances and longitudes.

—Roux de Rochelle,
"Mémoire sur la fixation d'un premier méridien," 1844

In February 1870 the distinguished astronomer and geodesist Otto Struve (1819–1905; see Figure 2.1) presented a study of the world's meridians to fellow members of the Geographical Society of Russia.[1] His address was in response to the plethora of initial meridians used in official and commercial cartography: on land maps of all sorts, coastal and hydrographic charts, world maps, atlases, and so forth.[2] For decades geographers and others had been complaining about European countries adopting their own national meridians,[3] a problem that had increased significantly as more nations erected astronomical observatories hard by their capitals.

Struve, whose own observatory—the Central Astronomical Observatory at Pulkova, near St. Petersburg—had been built in the early 1840s, focused on the issues associated with choosing one meridian to which all others would be referred. His position as director at Pulkova, which had become world famous for its authoritative role in determinations of meridian arcs,[4] and for linking the base longitudes of the astronomical observatories at Greenwich, at Altona in Denmark, and at Pulkova via

FIGURE 2.1

Otto W. Struve, director of the Pulkova Observatory, ca. 1870. Mary Lea Shane Archives of the Lick Observatory.

the Expédition Chronométrique (Grand Chronometrical Expedition), gave added authority to his remarks.[5]

Of course, the Russian astronomer was not the first to propose a unifying meridian. In addition to Pierre Laplace and Roux de Rochelle, whose views introduce this chapter, an 1868 technical report signed by Colonel Sir Henry James, director general of the Topographic Department of the British War Office, proposed a worldwide series of sectional maps whose first meridian would be that of the Royal Observatory at Greenwich. James did not explain the reason for this choice except to say that "[i]t is greatly to be desired that a first meridian and a uniform system for maps should be adopted for all nations, as the series of maps made in each country would then exactly correspond," and to assert that "as Greenwich is nearly in the centre of the habitable portion of the globe . . . the meridian of Greenwich should be adopted as the first meridian for all nations."[6]

Perhaps James's proposal was no more than a justification for a new map series already begun. He wrote that it was "too much to expect a concurrence of other nations in this [proposal]" and claimed that no real inconvenience would result if countries outside the British Empire continued to use their own base meridians. But for any successful map series along the lines James proposed, nations would have to agree upon a host of parameters—a common projection, latitude and longitude boundaries for the map sections, symbols, and legends—for the literally hundreds of sheets the program entailed. Even more challenging, changing a country's official base meridian would require at least a ministerial-level decision.

At this time fourteen separate base meridians were being used on European topographical maps.[7] So it is not surprising that only a small number of sheets envisioned in James's proposal were ever prepared, all of them by Great Britain's Ordnance Survey.[8]

In contrast to James, Struve offered a detailed study of the multiplicity of initial meridians (see Tables 2.1 and 6.2). His context was the damage to scientific efforts resulting from this expression of national pride and prejudices. Early in his remarks he reminded his audience of the great movement toward uniformity in measurement,[9] and the advantages that were likely to accrue, not only for science, but also for individual countries' industrial development. Cartographic uniformity was equally essential, Struve argued. Further, "The question of the unification of meridians does not depend on any consideration of political economy, it concerns solely the scientific world"—this view was often taken by scientists reluctant to address directly the extent of their country's feelings of pride and prestige.[10]

In his presentation Struve included a summary history of longitude scales on maps and charts beginning in the second century AD, Ptolemaic era.[11] With respect to the world stock of European cartography, he declared that three initial meridians predominated: the one passing through the island of Ferro (Hierro), westernmost of the Canary Islands, and defined as exactly 20° west of the meridian of Paris; the Paris meridian itself, the one passing through the Paris Observatory; and the meridian

TABLE 2.1

Initial meridians on nineteenth-century marine charts

Country	Meridian
Austria-Hungary	Greenwich
Belgium	Greenwich, Paris
Denmark	Greenwich
France	Paris
Germany	Greenwich
Great Britain & Ireland	Greenwich
Italy	Greenwich
Netherlands	Greenwich
Norway	Greenwich
Portugal	Lisbon
Russia	Greenwich, Pulkova
Spain	Cádiz
Sweden	Greenwich
United States	Greenwich

SOURCE: Data from U.K. War Office, *Government Surveys of the Principal Countries*, and (for Belgium) Jean-Pierre De Cuyper, private communication, 2004.

passing through the Royal Observatory at Greenwich (see Figure 2.2). The differences in longitude between the locations were sufficiently well known that, for all practical and most scientific purposes, adopting any one of the three as the world's initial meridian would be satisfactory.

Struve examined each of these three meridians in terms of their application in particular cases. Ferro remained the favorite on general maps for teaching purposes, thanks to the large numbers of elementary-school books being printed in Germany. However, in the case of maps and charts for scientific and practical uses, Struve stated that by far the most common initial meridian was Greenwich.[12] Moreover, as a reflection of the influence of Great Britain's vast output of reliable hydrographic charts embracing most regions of the world, two-thirds of the dozen European maritime countries already linked their own marine surveys to the Greenwich meridian.[13]

Struve turned to the specialist users of these maps and charts. Mariners,

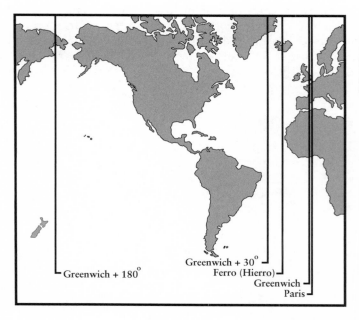

FIGURE 2.2

In his examination of candidates for a common initial meridian in cartography, navigation, and astronomy, Otto W. Struve considered the five shown here. His first choice was the Greenwich meridian, his second choice the Greenwich anti-meridian (i.e., 180 degrees from Greenwich).

geographers, and astronomers who needed to fix their location on the globe and/or investigate scientific questions were dependent on predictive tables listing the positions and movements of the stars at any given moment. According to him, the best of these ephemerides was the British *Nautical Almanac*, whose highly accurate tables were computed by astronomers and mathematicians and which was published by the Admiralty on an annual basis. He reminded his audience that in 1853 the Imperial Russian Navy had stopped publishing its own almanac as incomplete, inaccurate, and expensive to produce, in favor of the exemplary British one.[14]

Of the other ephemerides, Struve asserted that only the German and the American ones could be considered the British publication's peers in terms of precision and accuracy, noting at the same time that the German publication was less convenient for geographers. And although the *American Ephemeris and Nautical Almanac* did include tables based on the Naval Observatory's Washington meridian, its other tables were calculated for the meridian of Greenwich and so were essentially as convenient as the British tabulations.

Struve finished his analysis of the world's important ephemerides by criticizing those published by France, Italy, and Spain, which were based on their respective national meridians. Over the past years France's *Connaissance des temps*, for example, had fallen in astronomers' esteem, for the accuracy of the tables fell short of the urgent needs of science. He noted that Urbain Leverrier, director of the Paris Observatory, had prohibited the use of tables from the *Connaissance* at his institution.[15]

To Struve, the choice was clear: the Greenwich meridian should be the world's initial one. He ended his discourse by anticipating two objections from geographers in those countries unwilling, or unable, to relinquish their national meridians. First, the Greenwich meridian bisects Europe and Africa. As a result, longitudes linked to it must be reckoned as either positive or negative in different sectors of the two continents.[16] In contrast, those European countries whose maps have Ferro as the initial meridian show longitudes all of one sign—certainly a simpler notation.

As a matter of fact, this bisection issue had been specifically raised two decades earlier while the United States was debating choices for its own initial meridian.[17] However, considering the widespread use of British hydrographic charts as well as its *Nautical Almanac* among the world's navigators, errors arising from the consequent bisecting of longitude reckoning must be considered minor, as Struve indicated. Nonetheless, objections to expanding the use of a plus-or-minus correction certainly had some weight, for applying an incorrect sign in calculations is an all-too-common error, even for the most careful specialist.[18]

A second point in favor of the Ferro meridian was that it did not bisect any continent in its north–south sweep of the Atlantic Ocean. Consequently, Struve extended his analysis to consider other meridians that were exact hours (i.e., multiples of 15°) from Greenwich and shared this attribute. In his view, only two existed. One, two hours (30°) west of Greenwich, lies almost entirely in the Atlantic Ocean, cutting through only a small, nearly uninhabited portion of Greenland. The other, a Pacific Ocean meridian twelve hours (180°) from Greenwich, often called its "anti-meridian," passes through only a portion of the northeastern extreme of the Asian continent; it, too, was almost entirely uninhabited.[19] For Struve, both possibilities had advantages for science superior to those offered by Ferro.

The Greenwich anti-meridian in particular looks attractive because it lies in that region of the Pacific Ocean where navigators commonly add or subtract a day in their daily reckoning. As Struve would write later when the issue of worldwide timekeeping was being considered, choosing Greenwich-plus-180° as the initial meridian would mean that "the commencement of a new date would be identical with that of the hours of cosmopolitan time."[20] Moreover, were the anti-meridian selected, mariners' and geographers' use of the British *Nautical Almanac* (and all other ephemerides based on Greenwich) would be affected in the most minor of ways: merely the switching of midday to midnight in the tabulations.[21]

With the selection of three Greenwich-linked meridians, Struve's analysis was complete. Summing up, he declared that in his view, the one passing through the meridian telescope on the grounds of the Royal Observatory at Greenwich was the best choice from both a worldwide and a scientific standpoint.[22]

This 1870 study by a most distinguished scientist incorporated essentially all the arguments for choosing a single, initial meridian for navigation, geography, and astronomy. Within eighteen months its thrust was generally known to European geographers.[23] It can be considered the first salvo in what turned out to be a half century of skirmishes aimed at having the world adopt Greenwich as the common meridian for longitudes.

Three months after the treaty ending the Franco-German War was signed, the world's geographers assembled for the first-ever International Geographical Congress. Held in Antwerp in mid-August 1871, over five hundred participants from seventeen countries, including technical experts from the late belligerents, attended the week of sessions. Many came as delegates of regional geographical societies, a number of whom were employed by government mapping and charting organizations. No delegate came as an official government representative with power to make decisions affecting his country's cartography, however. The assembly was a purely professional and technical gathering.

The IGC had been established to discuss all interests and concerns associated with the science of geography, taken in its broadest sense. In the months preceding the congress a host of subjects had been raised formally, and additional issues were introduced at Antwerp. During the week there, delegates attended parallel sections ("groups") focused on areas of common interest and then met at general sessions to vote on specific questions.

Uniformity in cartography was an underlying theme, and seeking general agreement on a common initial meridian for cartography became the first item on the congress's agenda. Throughout the week it was discussed in the groups by those interested specifically in cosmography, navigation, and commerce and again addressed at the general sessions. The question of choosing a common meridian came before the delegates no fewer than six times—making it by far the most discussed of all the geographical issues raised.[24]

At the first combined-section meeting, participants noted the likelihood of error resulting from using different meridians at sea. They emphasized that the risk of error increased significantly when such values did not differ by very much, specifically citing the Greenwich and Paris meridians, which differed by only $2°20'$.

From the sole standpoint of navigation, Greenwich was the participants' unanimous choice: it was already used by the majority of sailors, and the *Nautical Almanac* was not only the most widely available publication, but was consistently published in advance of the other ephemerides.[25]

At one of the first general sessions Struve's 1870 lecture was briefly summarized, thereby bringing before the world's geographers his primary choice—Greenwich—as well as his two other proposals.[26] One participant argued that adopting Greenwich for all uses was not yet possible. Although Greenwich was the basis for the sailors' *Nautical Almanac,* many scientific studies were still linked to the Paris meridian. He recommended that the question of a common initial meridian be brought to the attention of government officials.

At this session, the majority of those attending voted in favor of adopting the Greenwich meridian. However, congress officials intervened, noting that other groups were considering the same issue and that a common position ought to be reached. After some wrangling, further exchanges regarding a common meridian were postponed.

On the last day of the congress, participants met in a general session to establish the opinion of the IGC regarding adoption of a common initial meridian. Before them was a proposed resolution, the result of numerous meetings, discussions, and negotiations:

The Congress expresses the position that for maritime routing charts [pilot charts], a first meridian be adopted, that of the Greenwich Observatory; and that after a period of time, say in ten to fifteen years, this initial position is to be made absolutely obligatory in the preparation of all charts of this nature.[27]

As a concession to those troubled by the apparent danger to navigation in the open seas inherent in using charts with differing base meridians, the resolution contained language agreeing that ships exchanging their respective longitudes should do so using Greenwich-based values exclusively. In reality, this meant merely that the geographers were encouraging a unifying action in someone's else's bailiwick. With regard to land maps and coastal charts, the resolution's position was that the advantages of each country's continuing to use its own base meridian outweighed the disadvantages.

Obviously the resolution was an extremely weak one. Official land maps and all sorts of privately published terrestrial maps—despite the multitude of base meridians they represented—were now excluded from consideration. Almost as limiting, the particular class of marine charts addressed in the resolution was a minor one. Given this lack of achievement, some participants in the drafting process felt constrained to express dissenting opinions. One of these, offered by the distinguished French economic geographer and historian, Émile Levasseur (1828–1911), is especially significant.

There are only two meridians to consider, Levasseur asserted: Paris and Greenwich. Paying his deep respect to a compatriot who had argued forcefully for Paris based on precedence, Levasseur insisted that a higher mandate was now in force. If one were living in the seventeenth or eighteenth century, it would have been quite proper to adopt the meridian of Paris to reflect France's dominant scientific position in geodesy and mapping. However, geographers and chart makers today should work wholly in the interest of mariners. Because the vast majority of charts currently in use were English ones, and because the Greenwich-based *Nautical Almanac* was accepted as the common navigational work (*le livre habitual du marine*), from a practical standpoint the choice must be the English meridian.[28]

Although Levasseur's remarks demonstrated an encouragingly transnational outlook, the official resolution of the IGC showed almost no movement toward increasing uniformity. Most hydrographic charts were already based on Greenwich; only those of France, Italy, Portugal, and Spain were still linked to different observatories.

Clearly, selecting an initial meridian to which countries would adhere was far too contentious an issue to resolve at this time. Nevertheless, it was absolutely necessary at this first congress to demonstrate that technical experts from different countries could work together and find common ground. So, despite its weaknesses and omissions, the delegates voted for the resolution as it stood. The First IGC was over.[29]

The issue of a prime meridian continued to fester, however. In May 1874, a translation of Struve's earlier report on the subject was communicated to the members of the Geographical Society of France and subsequently printed in its bulletin. It was immediately countered by the French hydrographer Adrien Germain (1837–1895).[30] Germain was secretary of the IGC's committee charged with drawing up the questions to be discussed at its next meeting. Thanks to his efforts, a majority of the committee had agreed to remove the subject of an initial meridian from the agenda. By articulating his reasons for doing so, he hoped when the congress met to win additional support for retaining the Paris Observatory meridian for all its current uses, including marine cartography.

Germain saw no reason for any country to adopt a neutral meridian, particularly some imaginary one not based on an actual, well-defined location on land. His examples included the somewhat ambiguous meridian of Ferro as well as that of Greenwich shifted by twelve hours, which traversed mostly ocean. He emphasized his opposition to such proposals by stating that if the day ever came for France to abandon the Paris Observatory meridian, it would be constrained to take that of Greenwich. Germain scoffed at the suggestion some proponents of unification had been advancing that if Britain were to adopt the metric system, it would be gracious for France to adopt the Greenwich meridian. He saw this as an example of scientific issues being trivialized by casting them as a mere exchange of gifts for maintaining friendly relations—political currency, as it were.[31]

Turning to ocean navigation, Germain saw only exaggeration in the claims of danger associated with the multiplicity of meridians on hydrographic charts, even when sailing in unknown waters. He reminded his audience that the *International Code of Signals*, a universal dictionary carried on all merchant and naval ships, had a particular code signal for requesting another ship's current longitude: FGH. Moreover, in the unlikely event of ambiguity, there was also the signal FCT, "What is your initial meridian?"[32]

The last part of Germain's paper was a detailed rebuttal of Struve's view that the British *Nautical Almanac* was the only ephemeris of sufficient quality for general use by navigators (and astronomers). Conceding that the Russian astronomer's downgrading of France's *Connaissance des temps* was the proper stance in 1870, he insisted that the *Connais-sance* had since undergone a massive revision and improvement, and that, in his view, it was once again every bit as good as the British annual volumes.[33]

In 1879 Sandford Fleming, a central figure in subsequent prime merid-ian debates, asserted that Germain's "line of argument . . . does not call for refutation"—thus dismissing the French hydrographer's views out of hand.[34] However, Germain's technically strong reasoning had already convinced many of his colleagues to continue using the Paris Observa-tory meridian on French hydrographic charts.

The second meeting of the IGC was held in Paris in August 1874. Even though the subject of an initial meridian had been removed as an official question, it remained an attractive topic for debate, producing several animated discussions during the ten days of sessions.

During the initial session of group II—Hydrography and Maritime Geography—participants considered numerous uniformity issues asso-ciated with the publication of hydrographic charts and other aids for mariners: notices, sailing directions, and the like. Early in his presenta-tion the French hydrographer Charles Ploix declared that for calculating longitudes, Greenwich was the only "meridian of origin" (*méridien de départ*), simply because of the vast number of marine charts based on it.[35] Of course, not everyone accepted Ploix's view. One participant re-minded attendees that the subject had been treated at the first congress, where members had voted that only routing charts need conform—and only after a period of time. Ploix's colleague Adrien Germain once again asserted that any inconveniences associated with using multiple initial meridians on sea charts were more apparent than real and, moreover, that "all intelligent captains" had no difficulty when longitude or time signals were exchanged; in other words, general adoption of Greenwich on the world's maritime charts was not needed.[36] Ploix, who was most anxious for the group to consider other uniformity issues associated with marine charting, conceded that from the standpoint of hydrography alone, the question of which meridian to use had little importance.

Undoubtedly remembering the lengthy discussions at the IGC's first congress, which resulted in almost nothing at all, and eager to move on to the other questions before them, participants decided that their group was not the proper venue for debating this question. Instead, they simply expressed the view that a common initial meridian ought to be adopted. The choice of which one should be left to an international scientific com-mission composed of accredited delegates from the world's maritime countries and charged with achieving uniformity in conventional signs on maritime charts. The congress adopted group II's view.[37]

Deferring to a yet-to-be-established international body was not only practical, it also recognized that an organization of individuals, no matter how professionally distinguished and capable, had no power to effect such a change. In the specific case of hydrographic charts, their preparation and maintenance were entirely the responsibility of national governments, whose hydrographic offices were almost exclusively staffed by naval officers.[38] Any far-reaching changes in these official products required either the approval of the respective ministry or some clear sanction by the country's legislature.

Selecting a common initial meridian also became a contentious issue during the last session of group I—Mathematical Geography, Geodesy, and Topography—during which attendees were considering projection schemes for geographical (small-scale) maps. At one point in his presentation, A. E. Béguyer de Chancourtois (1820–1886), a professor of geology at France's School of Mines, expressed a preference for a meridian 60° west of the ancient city of Alexandria (about 28°30' west of the Paris meridian), which passes through the Azores.[39] In de Chancourtois's view, this particular choice neatly separated the New World from the Old, and its sweep through the open Atlantic made its position an excellent place to alter the calendar day.[40]

During the ensuing discussion, one participant observed that the question of shifting the day was so important for the needs of education, international relations, meteorology, and telegraphy that its location had to be resolved promptly; he proposed that the meridian for changing the date be located in the Pacific Ocean, between the American and Asian continents. Not surprisingly, de Chancourtois responded that as for the meridian of origin, either Ferro or some other meridian passing through the Azores had greater merit.[41]

Another participant, the spokesman for a small group of members of the Geographical Society of Geneva, then proposed that the congress designate a north–south line through Jerusalem as the initial meridian, and that it recommend the use of this meridian by all countries once an astronomical observatory had been erected and was in operation. He argued that having an observatory sited on such "neutral ground" was most advantageous: its meridian would not pass through any nation's capital, and thus no country could possibly object to its selection.[42] Indeed, one does see in this proposal a reflection of the same transnational principles underlying the Convention of the Metre, which had been signed in Paris by official representatives of seventeen countries only a few months earlier.[43]

Pursuing the notion of a neutral initial meridian, another attendee proposed the adoption of one that would pass through the center of

Europe and Africa. However, others rejected the idea since its use would lead to plus and minus longitude values, urging instead that one of the ancient meridians to the west of Europe be adopted. Also in a neutral-meridian vein and consistent with his 1870 study, Otto Struve proposed as the world's zero meridian that of Greenwich augmented by 180°.

Perhaps other candidates were proffered at this session; its summary indicates only that the discussion continued, with many members taking part.[44] However, the proceedings did record the views of the French astronomer A. J. F. Yvon-Villarceau (1813–1883), who observed that a unique meridian was needed for general maps in future world atlases, while particular base meridians were indispensable in the case of local (i.e., large-scale) maps and charts. Then, in a portent of the future, he noted that the universal acceptance of a common initial meridian would permit the reporting of dates and times in terms of a common origin, though necessarily, of course, local times would continue to be used.

Yvon-Villarceau saw little hope for any near-term resolution of the initial meridian question. He articulated two unmet needs: (a) a precise set of longitude differences linking astronomical observatories situated in both the Old and the New Worlds, and (b) a large collection of carefully measured secondary values, repeated several times, to determine whether islands and continents actually preserved their relative positions. In his view, only after these data were in hand could one begin to consider choosing a common originating meridian.[45]

Nevertheless, all participants agreed that an initial meridian was needed in order to publish ensembles of land maps encompassing the world. (Not surprisingly, members also recorded their view that local maps—topographical map series—should continue to be based on specified local meridians, such as that of a country's national astronomical observatory.)

Those attending this long evening session of group I finally voted on the issue of a circumscribed initial meridian. By a large majority (eighteen to four) they declared that the meridian of the island of Ferro, which they defined as 20° west of Paris, should be used, adding that publishers of world atlases should specifically label this the "meridian of origin."

The Second IGC did not officially adopt group I's recommendation.[46] But the congress's discussion of various neutral meridians kept the subject open. So in the years following, a host of regional and minor geographical societies debated the possibilities.[47]

Perhaps the best-known non-Greenwich initial meridian was that proposed by Henry Bouthillier de Beaumont (1819–1898), an ardent student of geography and the founding president of the Geographical Society of Geneva.[48] In February 1876 he had called a special session of the Italian

Society of Geography, which was meeting in Rome, at which he presented his choice for a neutral initial meridian: one passing through the Bering Strait between Asia and North America.[49] The meridian's extension over the poles would be an arc passing through several European countries and the continent of Africa without touching any nation's capital—making it, in de Beaumont's view, a suitably international meridian. He proposed that the extension be called *médiateur* for its similarity to *equateur* and that longitudes east and west be counted from it.[50]

Over the next three years, de Beaumont pressed his choice at meetings of regional geographical societies and expanded his remarks, distributing published copies of them widely. His views were picked up in the popular press. In mid-January 1879 the French journal *L'Exploration* printed an extract of his most recent presentation, which was followed by favorable comments in the *Times* and *Nature*; in mid-February another favorable editorial appeared in the *New York Times*. All focused on the neutral meridian's passing through the Bering Strait, thereby missing the intent of de Beaumont's proposal.[51]

Despite the general tone of the publicity surrounding de Beaumont and his plan,[52] the world's geographers and other professionals had already spoken. Although a common initial meridian for hydrographic charts was most desirable, change could be effected only at a government-sanctioned conference. Large-scale maps, derived from national topographical surveys, would continue to be based on local meridians—most certainly until acceptably precise values of the longitude differences between central observatories were well in hand. Commercially published atlases would also continue as before, despite the view of many individuals that much greater uniformity in longitude scales was desirable.[53]

Indeed, a common initial meridian might not have remained an issue much longer, for the geographers' debates had degenerated into sessions focused on various proposals, each supported by some "important" criterion. However, the unexpected introduction of an entirely new consideration—a single time for special and general purposes—galvanized the process. Moreover, this new issue engaged a host of other technical experts, who met in their own international organizations.

CHAPTER 3

Enter Two Innovators

Two novel but related goals link the subjects of this section: to have North American railroads use one of four uniform times, and to adopt a single time throughout the world for specialized purposes. The first was articulated in 1869 by an American educator; the second came in 1878 from a Canadian railway engineer. After significant modifications, the first led to the inauguration of our worldwide system of Standard Time; the second is embodied in Coordinated Universal Time.

Charles Ferdinand Dowd (1825–1904), who graduated from Yale College in 1853 and was the owner and principal of a women's seminary in Saratoga Springs, New York, began his foray into timekeeping as an intellectual exercise (see Figure 3.1). It became a fifteen-year-long quest to have his system of timekeeping adopted by American and Canadian railroads.[1]

Since the dawn of railroading, every railway had selected its own time standard; by the late 1860s over eighty different ones were in use. The only rationale guiding the hundreds of companies was to select the local time of the city where they were headquartered and use it throughout their rail system. The civil times of towns and cities also varied markedly, but at least there was rhyme and reason to them, for they varied smoothly with east–west position. Indeed, by using a map, a traveler could determine the time difference between his own local time and that of any city. Not so for railroad times.

Dowd's initial concept was not very profound: have the railroads use a single time—the official one defined by the meridian of Washington, D.C.[2]—for their schedules; cities and towns would continue with the system of local times. The translation table between the two time systems would be published and sold as a gazetteer.

After receiving tentative approval for his concept from a group of railway companies, Dowd set to work, calculating the (local) time difference equivalent to the longitude difference between Washington and each of the more than eight hundred railroad stations being served by some five hundred railway lines. Since the United States is nearly four hours wide, east to west, he found differences of two and even three hours. Learning that the railway companies would not accept a timetable with such huge time differences, Dowd came up with the idea of having them adopt one of four times, each differing by exactly one hour from its neighbors. The set of times included Washington time; thus, the four time-defining meridians were spaced in multiples of fifteen degrees from the Washington meridian.

The publication in 1870 of Dowd's system, which he called National Time, with its map showing the United States and the lower part of Canada divided into four one-hour sections, was the first of its kind.[3] For the system to be a success, Dowd realized, it had to be not only endorsed by the country's railroad managers, but actually used by all the railways. So Dowd began attending the industry's regional meetings to present his ideas and obtain comments. He also printed other tracts summarizing his ideas.

Attempting to secure the railway companies' approval, Dowd subsequently switched the basis for his hour sections from the Washington meridian to the seventy-fifth meridian from Greenwich, better matching his plan to the country's rail lines. Yet this proposed shift (and another one to the New York City meridian) did nothing to advance his cause; and by the end of the 1870s, Dowd had faded into obscurity. Only the publicity

FIGURE 3.1

Charles F. Dowd, 1883. In 1870 the educator proposed the use of hour sections for North American railroads. Courtesy Skidmore College Archives.

associated with the railroads' 1883 adoption of Standard Railway Time made the general public aware of his attempts to create a uniform system of timekeeping.[4] Without doubt Dowd was the first to suggest a workable system of uniform timekeeping that could span a continent.[5]

Sandford Fleming (1827–1915), the other North American innovator, was born in Kirkcaldy, Scotland, and rose to great prominence in Canada. Trained as a surveyor, he arrived in his adopted country in 1845. Essentially all of his engineering work involved Canada's railways: surveying new routes, overseeing the construction of various rail lines, and supervising their subsequent maintenance. Fleming's professional career ended in May 1880 when he was forced to resign as engineer-in-chief of the then nascent Canadian Pacific Railway, his primary responsibility having been the government surveys for the line's route from Montreal to British Columbia. Elected in that year to the largely honorary post of chancellor of Queen's University in Kingston, Ontario (see Figure 3.2), Fleming spent the next two decades pursuing one of his passions: revising the tenets of public time.[6]

The oft-told story of the incident that first turned Sandford Fleming's attention to timekeeping is charming; it also has all the attributes of a myth. Briefly, in July 1876 the Canadian railway engineer made an overnight visit to a tiny village near the west coast of Ireland. The trip, which began in Londonderry, required a multiplicity of conveyances:

FIGURE 3.2

Railway engineer Sandford Fleming soon after his 1880 election as chancellor of Queen's University, Kingston, Ontario. Library and Archives Canada, C-14238.

railway carriage, horse-drawn public car, and private carriage. The trip down was uneventful. On the return leg Fleming arrived at a branch-line station expecting to board his train twenty minutes later—at 5:35 P.M. There he learned that his copy of the *Official Irish Travelling Guide* was in error, and that the train had already left—at 5:00 P.M. Fleming was forced to stay in the town until the next morning, returning to Londonderry almost a day late, thereby subjecting "several other persons to needless inconvenience and disappointment."[7]

Fleming's biographers have declared that while waiting to board the next day's return train, Fleming focused his thoughts for the very first time on reforming the world's timekeeping. A close examination of the incident leads inexorably to the conclusion that the typographical error in the Irish railway guide was not as Fleming stated in his first published pamphlet on timekeeping—that a 5:35 morning train had been labeled as leaving at 5:35 P.M.—but that he modified it to support his own ideas regarding time notation.[8] The long-held view that the missed-train incident led to Fleming's focus on time simply doesn't ring true. As an engineer heading the construction of his country's major railways, Fleming would have been intimately involved in timekeeping and the scheduling of rail operations, necessarily considering such issues as printed timetables years prior to an incident that occurred so late in his professional career. Indeed, one historian even suggests that in 1876 Fleming was already aware of Charles Dowd's 1870s attempts to alter timekeeping on North American railroads via the adoption of multiple time zones.[9]

Fleming's first biographer wrote that he began to write his first pamphlet, *Terrestrial Time*, almost immediately after the incident in Ireland and that the thirty-seven-page work was in print by the end of 1876—a span of no more than five months.[10] This assertion is incorrect, however. *Terrestrial Time* was not printed until the spring of 1878—a span of some twenty months, and a more realistic one, considering the work's content.

Whatever the true significance of the missed-train incident and its aftermath, sections of *Terrestrial Time* do propose a new approach to civil timekeeping.[11] Fleming's views eventually came to the attention of many professional groups, and *Terrestrial Time* is key to understanding his particular solutions.

It is also useful to compare Fleming's efforts with those of Charles Dowd a few years earlier. Not surprisingly, *Terrestrial Time* begins with the very same issue that was articulated by Dowd: the annoying and perplexing problem faced by a traveler on encountering the multiplicity of operating times used by railroads. Although Dowd focused on Americans traveling in the United States, Fleming emphasized the plight of a steamship passenger arriving in Halifax and then traveling west to

Toronto on Canadian railways.[12] Both wrote that cities and towns along
the routes kept their own particular local times and that these civil time
displays were basically different from the time shown on clocks at the
train stations. As Dowd wrote, "Any traveler . . . on leaving home, loses
all confidence in his watch, and is, in fact, without any reliable time,"
and Fleming echoed this view by noting that after arriving in a city, a
traveler is "obliged to alter his watch [to the locale's time] in order to
avoid much inconvenience and, perhaps, not a few disappointments and
annoyances to himself and others."[13]

In contrast to Dowd, Fleming also articulated a world view; his example was that of a traveler going from London to India (and on to China).
Starting with his watch set to Greenwich time, the traveler would encounter Paris time, Rome time, ship's time, Egyptian time, ship's time again,
and, on arriving at Bombay, two times: the city's local time and railway
(Madras) time. Had the traveler not altered his watch at all, it would now
be five hours slow. Had he missed one or two time shifts during his journey, he would be—to use Dowd's words—"without any reliable time."

Introducing a second difficulty, one not considered by Dowd, Fleming
noted the inconveniences and errors associated with dividing the civil
day into two twelve-hour periods, giving his troubles in the west of Ireland as the example. Then, after two textual excursions in his pamphlet
to set the stage, he turned to the now-linked problems of multiple times
and the division of the day.[14]

The railway engineer's overriding constraint was identical to Dowd's:
municipalities would necessarily continue their use of local time. His
solution also resembled Dowd's: to create a new system of timekeeping
for transportation and communications. However, Dowd's national time
encompassed only North American railroads, while Fleming's terrestrial
time extended around the world. Further, while Dowd sought to link his
national time for the railroads to local (civil) times via a gazetteer, Fleming opted for the simultaneous display of terrestrial time and local time
on all public and individual timepieces.

For constructing his time system, Fleming began with a mean solar day,
whose length he divided into twenty-four equal parts.[15] Each part was
exactly equal to the common (mean solar) hour, but Fleming argued that
they "ought not to be considered hours in the ordinary sense, but simply
twenty-fourth parts of the mean time" of one averaged revolution of the
Earth. To distinguish this new set of terrestrial hours from civil hours, he
assigned to each a unique letter of the alphabet (A through Y, dropping
J and Z).[16] In turn, he assigned each letter to one of twenty-four equally
spaced meridians on the surface of the globe, arbitrarily co-locating one
of his terrestrial-time meridians with the Greenwich meridian.[17]

To illustrate terrestrial time, Fleming introduced the concept of a standard chronometer within a skeleton terrestrial globe whose "terrestrial hour" hand would point to each of the lettered meridians in turn, returning to any particular one after twenty-four hours had passed (see Figure 3.3).[18] Not surprisingly, Fleming did not draw attention to the fact that his notional, twenty-four-hour device was different from the common timekeepers being manufactured. And though he repeatedly asserted that public clocks and individual watches could be modified to show both times "without in the least disturbing the machinery of the instrument," in fact they could not be so altered, for they had twelve-hour movements.[19]

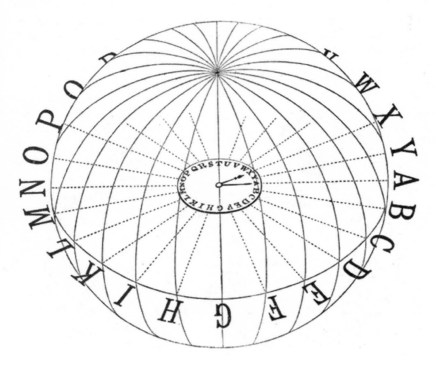

FIGURE 3.3

Sandford Fleming proposed the adoption of a single time, Terrestrial Time, to be used everywhere. To illustrate his proposal, he introduced the notion of a twenty-four-hour standard chronometer within a skeleton globe that represented the earth. Its "terrestrial hour" hand would move, pointing at local noon to one of the respective lettered meridians, and returning to its initially selected dial letter after twenty-four conventional hours had passed. From the cover of Uniform Non-local Time (Terrestrial Time), *n.p., [1878].*

Fleming's notation for a set of terrestrial-time meridians was most unusual. Apparently he was the first to use letters in place of the hours, minutes, and seconds often used by specialists when giving the longitude of a meridian in terms of time. Further, he argued that telling time by numerals had "no special advantage over letters." Were the naming of the twenty-four divisions of the day able to be done afresh with letters, "There can be no doubt whatever, that the time of day could be as well expressed, and be as easily understood," as it could using numerals.[20] It would take several iterations over the next years before Fleming finally discarded lettered hours—and not before he had had fabricated twenty-four-hour watches with letter-hour dials (see Figure 3.4).[21]

To demonstrate how terrestrial time would be used by steamship and railway companies, Fleming provided paired sets of timetables. One was a Cork-to-London route table; the other was a timetable for the same route, given in terms of the new system to demonstrate how it would

FIGURE 3.4

*For his system of timekeeping, Fleming designed a twenty-four-hour timepiece
with hours corresponding to the lettered meridians of his plan. In this first
version the Greenwich meridian is designated "G." When the watch is open, its
hands point to N.03 (the hour hand is the longer one). When the case is closed,
an adjustable dial attached to it comes into play, and in this example shows 6:03
P.M. Fleming also shaded the dial to show the sailors' four-hour duty watches.
From* Uniform Non-local Time (Terrestrial Time), *n.p., [1878], figures 6 and 7.*

remove the need to print multiple railway times. The second set, a theoretical one, went from London to Canada's Pacific Coast via steamship
and rail lines, including some not-yet-constructed ones. Most unfortunately, both sets are replete with time-conversion errors.[22] Contemporary
readers striving to understand Fleming's radically different time notation
must have thrown up their hands in dismay at finding themselves unable
to reconcile the entries.

Terrestrial Time's saving grace is a single paragraph in which Fleming
suggested that "each of the twenty-four lettered meridians . . . should be
taken as standard longitudes for establishing approximate local time,
and that as a general rule all places should adopt the local time of the
nearest of these meridians."[23] The result would be the division of the
globe into twenty-four "distinct local time sections"; within each all
would use the same time. This suggestion, of course, is for a worldwide
system of time zones.[24]

Fleming summarized this concept via two dozen illustrations of clock
faces. On all of them the hands depicting terrestrial time were fixed at
G.45 (12:45 P.M. Greenwich time) while an outer, double-twelve hour
dial was shown rotated to one of twenty-four positions to provide the
necessary link to approximate local time. Alongside each illustration
Fleming listed specific places whose local times approximated the time
at the respective lettered meridian. Here was visual proof that a uniform
time system linked to a single time could be used everywhere on earth—
albeit via an unusually complex notation. In Fleming's view, terrestrial
time was the key to all public timekeeping, to which local times were
subservient. This latter idea was, in my judgment, a major detriment to
general acceptance and understanding.

In the final section of the pamphlet, Fleming discussed the difficulty
of assigning a precise date to an event because the current system of local time reckoning meant that parts of three days existed simultaneously
around the globe.[25] To address the problem, he proposed the designation of a terrestrial-time meridian as an "initial meridian to denote the
dividing line between each day," one situated "through or near Behring's
Straits, passing from pole to pole through the Pacific Ocean, so as to
avoid all Continents and Islands."[26] Fleming would extend his remarks
on this issue in his next work.

If *Terrestrial Time* and its variants had been Fleming's sole contribution, they would probably be cited only as a footnote, particularly
since his two attempts to bring them before professional groups both
failed.[27] Moreover, others—Struve with his Greenwich-linked meridians and Dowd with his hour-difference sections—anticipated many of
Fleming's ideas. Privately published and complex in its exposition, his

work addressed a problem whose importance even he acknowledged lay mostly in the future. The pamphlet's focus is almost exclusively on time itself; there is little mention of time's link to longitude, and the need for a common initial meridian in geography is only alluded to.[28]

Rather different was the reception accorded Fleming's next publications, "Time-Reckoning" and "Longitude and Time-Reckoning," which appeared together early in 1879.[29] Although the complex framework associated with terrestrial time—now called cosmopolitan time—remained, and *Terrestrial Time*'s text was incorporated essentially in toto, Fleming's addition of longitude considerations and the recognition of the problem of multiple initial meridians in his second paper made these works a useful discourse.[30] Clearly, Fleming had become aware of the debates associated with the subject of a common initial meridian for geography and cartography and had incorporated them in his thinking.[31]

In "Time-Reckoning" Fleming begins by noting the ambiguity of time and date resulting from the lack of a common meridian. This was a well-known problem and had been mentioned in passing by the French astronomer Yvon-Villarceau at the Second International Geographical Congress.[32] Illustrating this topic with a repeating world map showing that under the current system of local times portions of three days existed concurrently, Fleming made obvious the resulting inconvenience. Reminding his audience that even people living along the same meridian can have different days as a result of the difference between Eastern and Western dating on Pacific Ocean islands, Fleming elaborated on the resulting confusion.[33]

In this paper, Fleming also expanded on his earlier notion of approximate local times: for all public purposes within a region, use the closer of the two adjacent time-defining meridians. The United Kingdom of Great Britain and Ireland was his example of a country where single-meridian times had replaced local times: Dublin time for all of Ireland and Greenwich time for England, Scotland, and Wales.[34] He also added a column of approximate local times to his previous London–Pacific Coast timetable, making them understandable by showing them with a numerical notation: London's 8:00 P.M., for example, was now shown as 20.00.[35]

Anticipating his next paper, Fleming wrote that "under the system of cosmopolitan time, the meridian which corresponds with zero would practically become the initial or prime meridian of the globe." To each of his letter-designated meridians Fleming also assigned a numerical longitude value, starting with 0° and increasing in 15° increments from east to west around the globe, ending at the start point with 360°.[36]

In "Longitude and Time-Reckoning," the shorter of the two papers,

Fleming makes clear his initial meridian choice: the Greenwich anti-meridian—which he defines as 0° (and 360°), and to which, in terms of cosmopolitan time reckoning, he assigns the letter Y. The paper's accompanying illustration represents the first integration of Fleming's time and longitude proposals.[37]

It must be noted that Fleming employed almost the same framework that Otto Struve used in his 1870 lecture to the Geographical Society of Russia: enumerating both historical and current meridians and then searching for meridians that, by not passing through any populated region on the globe, could be considered common to all nations. Fleming's desire to have his meridian for longitude be the one at which the date changed and his enumeration of its additional advantages with respect to Greenwich's overwhelming use on ocean navigation charts and the supremacy of the British *Nautical Almanac* both echoed Struve's comments regarding the Greenwich anti-meridian.[38]

Quite original with Fleming, however, and a rather clever way to demonstrate the importance of Britain's maritime products, was the inclusion of a table, "prepared from the latest authorities within reach," linking various nations' shipping—by number of ships and by tonnage—with the initial meridian they currently used. Not surprisingly, Greenwich led all the rest, with France's Paris Observatory meridian running a distant second.[39] Fleming's approach and his summaries were often quoted in later debates by those favoring the Greenwich meridian.

Fleming predicted that his choice of the meridian 180° from Greenwich, when eventually adopted, would cause the expression "the longitude of Greenwich" to fade away. He argued his choice was "essentially cosmopolitan"—neutral—in character and ended the formal part of his paper by declaring:

As the line of demarcation between one date and another it would be of universal interest, and a property common to the hundreds of millions who live on land, and the hundreds of thousands who sail on the sea.[40]

In this 1879 exposition, the railway engineer did not acknowledge Struve's earlier remarks regarding the anti-meridian of Greenwich as a possible choice; however, he did lavish praise on the Russian astronomer's (translated) paper, calling it "valuable" and worthy of perusal.

As already noted, in his summary remarks Fleming dismissed the French hydrographer Adrien Germain's 1875 rebuttal of Struve's analysis, terming it a "national and non-cosmopolitan view" of the subject of a prime meridian. While correct in this characterization, Fleming probably did not appreciate how important it had been in stiffening resistance to any meridian change on French hydrographic charts.[41] However, he probably wouldn't have cared. He was not writing an objective analysis of the need for a new initial meridian, nor even really couching his arguments in terms of selecting a neutral one. Rather, he had already declared the need for change and was proselytizing for a Greenwich-linked choice.

That Fleming was considering some sort of campaign to gain the adoption of the Greenwich anti-meridian can be seen in his response to de Beaumont's 1876 Bering Strait meridian. He seized on it, quoting verbatim its recent summary in the *Times Weekly Edition* and asserting that because it had been arrived at independently, it confirmed, even supported, his views. Though de Beaumont's preferred meridian (168° west of Greenwich) cut through Alaska's Cape Prince of Wales on the North American continent, while his own 180°-from-Greenwich passed through Asiatic Russia, Fleming proselytized that any difference between the two proposals was only in "matters of detail." One such "detail" was a time difference between the two meridians of forty-eight minutes![42]

But with the publication of the two papers in the *Proceedings of the Canadian Institute*, Sandford Fleming was now armed with something more useful than an obscure, privately printed exposition.

CHAPTER 4

Ventilating the Issues

It appears desirable that the question should be extensively ventilated by the memorialists.
—George B. Airy to the secretary of state for the colonies,
18 June 1879

Most of the issues associated with the now linked subjects of a universal
time and a common initial meridian were considered again and again.
Actions took place primarily within professional organizations, but with
only one or two members actually pursuing change. However, these ac-
tivists also tried very hard to enlist the support of national governments
and eventually achieved some success.

As of 1879 almost nothing had changed in either Europe or North
America. The limited recommendations of geographers had been ignored, so the world's hydrographical charts and topographical maps
continued as before, their basis particular countries' meridians. The only
change of note was the recommendation by an international congress of
European meteorologists that a common initial meridian be adopted for
synoptic maps, and, further, that it should be the Greenwich meridian.
Again, implementation required decisions by governments.[1]

In the area of public time, many different "systems" existed. In the
United States and elsewhere a miscellany of railroad operating times
existed, times often used by the small cities and towns hard by their
respective rail lines. In large cities such as Chicago, Toronto, New
York, Edinburgh, Marseilles, and Berlin, a single time extended across
the entire city, one based on either the meridian bisecting the building
that housed the seat of government or the meridian of a local astronomical observatory. In some countries a national time based on the
capital's meridian was also embraced. Everywhere else, local time was
the norm.

However, a completely unexpected problem appeared in the United States. In 1875 a sometime astronomer and government meteorologist-geophysicist named Cleveland Abbe (1838–1916) found that he was unable to place a large set of observations of the northern lights on a consistent time base. After realizing that some respondents were using railroad time in place of their town's local time, he wrote the New York–based AMS, a group of educators and scientists whose members debated various uniformity issues.[2] Abbe urged it to take action "to secure the adoption of a uniform standard of time." In response, the society established the Committee on Standard Time and made Abbe its chairman (see Figure 4.1).

In May 1879 the society finally approved for release its "Report on Standard Time," whose major themes were accuracy and uniformity in timekeeping.[3] Unquestionably the key document leading to standard time in the United States, the report recommended that North American railroads and telegraph companies adopt the time defined by one of five regional meridians spaced exactly an hour apart and linked to the Greenwich meridian. The document further recommended that all cities and towns drop their own local time and use instead the local time represented by one of the meridians. Although almost entirely focused on North America, the report included the committee members' opinion that the meridian 180° from Greenwich was convenient and the most practical as the basis for a uniform standard of time for the entire world. This endorsement was tempered by the concession that the question of

FIGURE 4.1

Cleveland Abbe, ca. 1874. The astronomer and geophysicist's maneuvering helped gain the passage of legislation for the 1884 International Meridian Conference in Washington. Mary Lea Shane Archives of the Lick Observatory.

worldwide uniform time was far in the future and "should be considered in some international convention." Printed copies of the "Report on Standard Time" became available the following March.[4]

Also in May 1879, the Marquis of Lorne, governor-general of Canada, sent to the secretary of state for the colonies a pamphlet containing Sandford Fleming's two recent papers and a memorial from the Council of the Canadian Institute. Approaching the governor-general had been a crucial strategic move on the part of the Canadian advocates, for his personal links to the crown ensured that his communications would receive careful attention from the home government's officials.[5]

The Canadian Institute's memorial asserted that Fleming's position as chief engineer of the Canadian Pacific Railway lent credence to his suggestion for a first or prime meridian, and that his choice was "free from the objections hitherto urged against other propositions" and so offered "an acceptable solution of a problem of international importance." It further noted the "peculiarly favorable position" occupied by the governor-general of a dominion stretching from the Atlantic to the Pacific Ocean, which allowed him "special facilities for promoting the simplification of a complex [current] system." It concluded with the additional assertion that Fleming's solution was not only extremely simple but also "free from the sources of international jealousy which have hitherto neutralized the efforts of scientific men to remedy practical evils which are universally recognized."[6] Although we may smile today at the memorial's exaggerations, in sending it the leaders of a professional institution had taken a significant step: enlisting the offices of a national government to further both uniform time and the adoption of an initial meridian.[7]

After these various documents reached London, the secretary of state for the colonies sent them to the government's principal science adviser, George B. Airy (1801–1892; see Figure 4.2). One of the kingdom's most distinguished astronomers royal, Airy was also director of the Royal Observatory at Greenwich, which was under the aegis of the Admiralty. In his reply to the secretary of state, Airy tendered a most negative assessment of Fleming's two proposals.

Airy argued that any degree of confusion caused by the use of local times varying by east–west positions had already been solved. As one example, he cited the railway station at Basle, Switzerland. Here the French and German railways met, and each system's clock displayed a different time (and, both, of course, were different from Basle's local time). Yet, he argued, there was no confusion or difficulty. Citing Fleming's case of a railway traveler journeying from New York to San Francisco, Airy opined that the solution was to run westbound trains on New York time and eastbound ones on San Francisco time, with double clocks at every

FIGURE 4.2

Astronomer Royal George Airy, ca. 1880. Airy was dismissive of Sandford Fleming's 1879 proposals sent to the British government. However, he suggested they be considered by private-sector organizations. Mary Lea Shane Archives of the Lick Observatory.

station. At the end of his journey, the traveler would simply alter his watch to conform with the particular locale's local time.

Airy reminded the secretary of state of the twenty-five-minute time difference between England and Ireland, which was not the result of legislation but which, he felt, was simply "the result of common sense, guided perhaps by a single personal influence."[8] He dismissed the proposal for cosmopolitan time and its associated lettered local-time standards, concluding, "I set not the slightest value on the remarks extending through the early parts of Mr. Fleming's paper."

Turning to Fleming's second proposal, a prime meridian for all nations, surprisingly Airy did not criticize the railway engineer's specific choice—the one located 180° from the Greenwich meridian—even though its selection would require the Admiralty to change the numbering of the longitude lines on its three thousand–plus ocean charts as well as to reformat and recalculate the various navigation tables and instructions for its *Nautical Almanac*. Instead, the astronomer royal insisted that "no practical man [i.e., navigator] ever wants such a thing" as a prime meridian.

Continuing in this rather rhetorical manner, Airy asserted that if a prime meridian were ever adopted, it would have to be Greenwich, for "the navigation of almost the whole world depends on calculations founded on that of Greenwich." After noting that most navigation was based on the *Nautical Almanac* and estimating its annual sales at more than 32,000 copies, he repudiated the idea of basing any claim to adopt it simply on the tradition of Royal Observatory calculations' reliability or on the number of almanacs sold. Perhaps remembering the loss in quality and resulting low esteem of the *Connaissance des temps* in times past and the somewhat checkered history of the British almanac itself,[9] Airy closed with a meridian selection based on technical considerations:

Let Greenwich do her best to maintain her high position in administering to the longitude of the world, and Nautical Almanacs do their best, and we will unite our efforts without special claim to the fictional honour of a Prime Meridian.[10]

The secretary of state had asked for suggestions on how to deal with the governor-general of Canada's request for possible government action. Airy, noting that the memorialists wanted the imperial government to bring the matter to the attention of officials and scientific groups in Great Britain and other countries, added that he could not imagine it taking on this responsibility. Thus, in his final paragraph Airy proposed as a response

[t]hat Her Majesty's Government, recognizing in some degree the inconveniences described by the memorialists, are not able to compare them with the possible inconvenience which might arise from the interference of the Government in such a matter. That it has been the custom of Her Majesty's Government to abstain from interfering to introduce novelties in any question of social usage, until the spontaneous rise of such novelties has become so extensive as to make it desirable that regulations should be sanctioned by superior authority. That it does not appear that such extensive spontaneous call in reference to the subjects of the Memorial has yet arisen. That it appears desirable that the question should be extensively ventilated by the memorialists and should be submitted by them to the principal geographical and Hydrographical bodies, including (perhaps with others) the Royal Geographical Society, and the Dock Trustees or other commercial bodies, in London, Liverpool and Glasgow.[11]

In my opinion, Airy gave the secretary of state a reasonable response to some novel, untested ideas. However, other historians have faulted the astronomer royal, blaming his opposition to a common initial meridian on shortsightedness and citing his successor's warm support as evidence that Airy was a soon-to-retire, out-of-touch government administrator.

But the astronomer royal had rejected not one but *two* proposals: a new way of displaying time and a prime meridian. There is no question that the idea of telling time with letters was far too radical for him—or the general public—to embrace. With regard to a common meridian,

Airy was well aware of the conflict among geographers regarding its selection, and even though Fleming's proposal was billed as a "neutral" solution, it was obviously not. More to the point, why should the British government's chief science adviser even consider a shift in base meridian for the huge stock of hydrographic charts used by most of the world's navigators, much less eagerly support it? Finally, Airy's successor as astronomer royal, William Christie, was to argue in support of the Greenwich meridian—not its anti-meridian—and he joined this fray only *after* Greenwich had already been accepted by European geodesists.

The ink on Airy's response had barely dried when the Canadian Institute began the second phase of its campaign by sending 150 copies of Fleming's bundled papers to the governor-general of Canada.[12] Included was a request that these be sent to scientific societies and to the representatives of foreign governments in London. In his official transmittal to the secretary of state, the governor-general attached the Canadian Institute's list of proposed addressees in eighteen countries, which included seven British groups. The astronomer royal's view that Her Majesty's Government should not be involved in any distribution process had been trumped.

There was very little initial support in Great Britain for Fleming's proposals.[13] Among official responses, the Admiralty responded negatively, as did the astronomer royal for Scotland. The Council of the Royal Astronomical Society met and decided to take no action in the matter; and the proposals were reviewed by the president of the Royal Geographical Society, who termed them impractical. Only the Royal Society made slightly favorable comments, judging, however, that the proposal for cosmopolitan time stood a greater chance of being adopted than that of a common prime meridian, for the latter involved the "susceptibilities of individual nations." It also warned that no such scheme would have much chance of success until "a general readiness on the part of civilized nations seriously to entertain the question" existed.[14]

The Canadians learned of the imperial government's position and the generally negative views expressed by British organizations late in 1879. The secretary of state for the colonies also emphasized that "Her Majesty's Government are merely transmitting these papers [to the various organizations] out of courtesy to a Scientific Institution in Canada, and that in doing so, they lend no support to, and assume no responsibility for the views advocated therein."[15] Given this official position, only Airy's suggestion that the subjects be publicized by the memorialists offered the Canadians any hope at all. They did nothing further until April 1880, when the AMS's "Report on Standard Time" was about to appear in print.

A month or two prior to the report's appearance, Sandford Fleming and Cleveland Abbe began to correspond. Undoubtedly seeking an ally, Fleming was aware of the latter's work on timekeeping via his earlier interaction with the AMS's president, F. A. P. Barnard.[16] Fleming and Abbe exchanged publications, and Abbe outlined the society's strategy: first to have telegraph and railroad companies set an example by adopting an operating time based on the Greenwich meridian, and then to "hammer away at our national congress and call for its action on the subject." Abbe also suggested that either the Canadian government or the governor-general call for an international convention, feeling that although the American government would be quite willing to appoint a representative, it would be most difficult to convince it to take the initiative.[17] Events would prove otherwise.

In April the Canadian Institute prepared a long memorandum in which it emphasized both the need for time uniformity in Canadian and American railroad operations and general scientific interest in time uniformity throughout the world. Besides citing Fleming's papers and the AMS's efforts and report, the memorandum quoted from the only semi-favorable response available, the comments of the Royal Society. During the next month, copies of Fleming's bundled papers and Abbe's report were sent by Canada's governor-general to eleven scientific societies on the European continent along with a request for their opinions on the two issues. Apparently only two replies were received, one from St. Petersburg and one from Berlin—certainly a most disappointing response. Both were quite favorable, however, stemming from the 1870 analysis by Otto Struve.[18]

In 1880, Sandford Fleming's life altered dramatically. Early in the year he stood for election as chancellor of Queen's University in Kingston, an honorary position that carried very light duties. He was victorious, and in May his link to the still unfinished Canadian Pacific Railway was severed.[19] The now retired but comfortably situated civil engineer turned his attention more and more to timekeeping issues.

Fleming became a member of the AMS. He continued to interact with Barnard and Abbe, who urged on him a much less radical notation than letters for the twenty-four cosmopolitan time meridians. They also suggested that the time itself be called "Greenwich time." Fleming's later papers indicate that he heeded their advice regarding notation and adopted specific suggestions regarding accurate time signals and their distribution addressed by Abbe. In short, a bond was formed between the Canadian Institute and the AMS, with Fleming as the glue.[20]

Fleming also joined the ASCE, which met in Montreal in June 1881. There he presented a paper, "On Uniform Standard Time, For Railways,

Telegraphs and Civil Purposes Generally." This led to the creation of the Special Committee on Standard Time, with Fleming as chairman, which became an important voice in subsequent lobbying for uniform time.[21]

In August 1881 Barnard floated Fleming's idea of a worldwide time system at the annual meeting of the Association for the Reform and Codification of the Law of Nations, but the results were discouraging.[22] The following month the International Geographical Congress held its third meeting in Venice, with more than a thousand individuals registered for a week of sessions.[23] The AMS and the Canadian Institute both sent delegates, as did seventy-nine other societies, thirty-six of which were geographical ones.

Several attendees spoke on the subject of an initial meridian, all before group I, Mathematical Geography, Geodesy, and Topography. Among the North Americans who spoke were Sandford Fleming, representing both the AMS and the Canadian Institute, and Captain George Wheeler of the U.S. Army Corps of Engineers, a commissioner and official delegate of the United States. Fleming presented remarks on the adoption of a prime meridian; Wheeler presented a communication prepared by Cleveland Abbe for Brigadier General William B. Hazen, chief signal officer of the U.S. army; and the president of the American Geographical Society presented a statement from the president of the AMS, F. A. P. Barnard, based on Fleming's ideas.[24]

These communications and numerous associated resolutions focused on the Greenwich anti-meridian for both longitude and uniform time, a choice that attendees were unwilling to accept. Some participants reminded their fellows of the resolutions adopted at the Second International Congress, while others continued to argue for Ferro or de Beaumont's Bering Strait projection.[25] Suggestions of different letters and/or names for the meridians of Fleming's cosmopolitan time led to additional discussions.

Together the American and Canadian delegates proposed that an international conference be held in Washington in May 1882 and indicated Hazen's willingness to form an executive committee from their ranks in order to make the necessary initial preparations, thus signaling interest by an agency of the U.S. government.[26] Though united, the North Americans did not prevail, but they salvaged their efforts by accepting a much watered-down resolution:

Group I expresses the hope that within a year Governments appoint an international commission for the purpose of considering the subject of an initial meridian, taking into account not only the question of longitude, but especially that of hours and dates. . . . The President of the Italian Geographical Society is requested to undertake the steps necessary to realize this view via his Government and the foreign geographical societies.[27]

Because the resolution was not taken before the entire congress, it represented merely the views of group I. Thus the resolutions adopted at previous congresses remained in force.

About a year later, the Italian government began distributing a circular letter prepared by the Italian Geographical Society, a step that did not commit it to any action and was reminiscent of the British government's quite modest response with regard to Fleming's materials. The letter asked the foreign societies that had been represented at the congress for their views on convening an international commission. Approximately ten replies were received, many of them favorable to the idea. However, the most forceful response was a negative one from the Royal Geographical Society in London. Noting the widespread use of Greenwich-based nautical charts, particularly by the British navy, the society stated formally that it could not take part in any conference directed toward the adoption of a different initial meridian.[28]

As we shall see, the Italian Geographical Society's circular was distributed rather too late, for the American government had already taken official action in the matter. Nevertheless, this tardy European effort, initiated by a tentative resolution at the Third International Geographical Congress, had one important consequence: it caused the International Geodetic Association to turn its attention to the joint issues of a common initial meridian and universal time. When the association deliberated in Rome in October 1883, its discussions and recommendations finally cleared the air.

CHAPTER 5

North America and Rome

Thus far, attempts to have professional societies pressure governments to take action on a common initial meridian and a universal time had met with no success. Undoubtedly Fleming and the American delegates returning home in the fall of 1881 from the Third International Geographical Congress were extremely disappointed by the lack of progress abroad. Efforts shifted to the United States and Canada.

Already astronomers had responded to the availability of Cleveland Abbe's seminal "Report on Standard Time." At the annual meeting of the AAAS in Boston in August 1880, those astronomers selling time signals established the Committee on Standard Time and began working with Abbe and the AMS to convince the country's railroad and telegraph companies to adopt one of the four 1-hour-difference times for their operations. Late in 1881 the chairman of the committee sent a letter to the railroad industry's General Time Convention, soon to become the key organization in the process that would lead to near-uniform time in the United States.[1]

Sandford Fleming, who in mid-1881 had championed the establishment of the Special Committee on Standard Time by the ASCE, now turned his attention to the committee.[2] He prepared a questionnaire comprising eleven "Questions Relating to Standard Time." It emphasized timekeeping in North America and included practically no mention of a geographical initial meridian, only suggesting in a companion piece included with the questionnaire that a "time zero" meridian be co-located with the geographic one somewhere in the Pacific Ocean.[3]

Fleming's questionnaire was approved for distribution by the ASCE's governing body in February 1882 and mailed to a select group of railroad and telegraph managers, astronomers, and civil engineers. Replies were received from 22 individuals in Canada, 114 in the United States, and 1 in Mexico. The respondents were almost unanimous in their support for time uniformity throughout North America, with three-quarters of them favoring time standards differing by exact hours.[4]

At the ASCE's May meeting in Washington, the Special Committee on Standard Time met to discuss the results of the survey and recommend a course of action. With the events taking place in Congress, the committee now proposed a petition to Congress urging it to take the necessary steps to establish a common initial meridian for longitude, to be agreed upon first by both Europeans and North Americans. It also recommended that if efforts to gain agreement failed, then "the people of the Western Continent should determine a zero meridian for their own use and guidance, with a view of establishing, as speedily as possible, a suitable time system for the United States, Canada and Mexico."[5] The officers of the ASCE accepted the committee's suggestions, and eventually a petition was sent by the society to Congress, joining others already before it.

One of the appeals for government intervention had come from the AMS. As we have seen, in his 1880 letter to Fleming, Abbe had articulated the need for a government-sponsored international convention to achieve widespread time uniformity. Most likely he soon learned of Canadian officials' inability to do anything in the matter. So, starting early in 1881, and with the vigorous support of his direct superior, General William Hazen, Abbe took steps to make the U.S. Signal Office the center for uniform time activities in the United States. As chairman of the AMS's Committee on Standard Time, he was involved with its decision to send a memorial to Congress urging an international convention.[6] Before the end of the year the process had been initiated, and the society sent a petition to various individuals with a request for their signatures.[7]

Suddenly a critical new issue appeared. A bill assigning the Naval Observatory responsibility for distributing Washington time signals throughout the country was reintroduced in the House of Representatives. Stung by this development, Abbe responded by publicly criticizing the Naval Observatory's time service. The ensuing messy fight between the U.S. Signal Service and the U.S. Naval Observatory raised serious questions about whether the Signal Service should have any role in public timekeeping. This squabble did not deflect Abbe and Hazen, who continued their efforts toward the establishment of a uniform time; however, their agency's activities were now under close scrutiny.[8]

The petitions to Congress for an international conference to consider a common zero meridian for longitude and time reckoning and the bill supporting the Naval Observatory's role placed two overlapping and conflicting proposals before both Houses. Throughout the early months of 1882 the legislative conflict created an air of uncertainty. Indeed, only in the last days of the congressional session did the bill to expand Naval Observatory activities in national time services fail and a joint resolution giving the president authority to call an international conference pass. It was 3 August 1882 when President Chester A. Arthur approved this legislation, and its proponents could finally breathe a sigh of relief. A few months later the secretary of state queried several countries on behalf of the president as to the desirability of holding such a conference.[9]

The existence of the now-approved joint resolution and Sandford Fleming's ongoing efforts within the ASCE affected the AAAS. In mid-August, at its annual meeting in Montreal, it established a second time-related committee—the Committee on a Primary Meridian and an International Standard of Time—to confer with delegates to the international convention once they were appointed. This committee also considered Fleming's printed *Letter to the President of the AAAS* (the now-chancellor of Queen's University having just joined the AAAS, undoubtedly for the express purpose of presenting his missive).[10] In this letter, Fleming urged that a separate convention be held whose delegates would represent "Scientific Societies, Railway Corporations, Chambers of Commerce, Departments of State, and other bodies throughout the United States, Canada and Mexico." These delegates would examine all issues associated with time uniformity and then "determine and recommend a system for regulating time which will secure the greatest advantages to all interested, in every locality in North America."[11]

In his letter, Fleming also stated that the North American convention he was proposing should take place after worldwide agreement on a prime meridian had been reached. All the while, however, he was arguing at the ASCE that Canada and the United States should take immediate steps to reform their own timekeeping via such a convention, asserting that the need to do so had become urgent.[12] His sole piece of evidence was the response to his own survey questionnaire, indicating the respondents' general approval of particular choices—*not*, however, agreement on the need to change immediately to a new system of public timekeeping. Although the ASCE supported Fleming's views, it moved slowly.[13]

It must be noted that Fleming's position in 1882 was that of chancellor of a university, largely an honorary post. No longer able to directly influence the highest levels of Canada's government, his forums were now professional societies, organizations with well-established procedures for approving actions taken in their names. Partly as a result of their slow pace, Fleming's influence waned, for events in the United States began to move much more rapidly than anyone could have foreseen—thanks to a railroad man named William F. Allen (1846–1915; see Figure 5.1).[14]

FIGURE 5.1

Railway journal editor and association secretary William F. Allen convinced the North American companies to adopt Standard Railway Time in November 1883. Engraving from Frank Leslie's Popular Monthly *17 (April 1884): 385.*

As part of his efforts to have the United States adopt a uniform system of timekeeping, Cleveland Abbe had urged the AMS to invite a number of railroad and telegraph officials, Allen among them, to become members so they could assist in attaining the goal. But Allen, a railway engineer by training, editor of the *Travelers' Official Guide* (a monthly compendium of route schedules and railway company organizations), and secretary of the railroad industry's General Time Convention (a group set up to harmonize train schedules), was never informed of his election to membership. He first learned of it in December 1881, and then only by accident.

Allen must have been astonished by the discussions at the late-December meeting of the AMS he attended. For the first time he learned of the many actions that influential people, in no way connected with railroads, had taken to promote uniform time. Among these activities was the planning then underway to involve the federal government in an international conference on worldwide time standards. At the meeting, he learned that a bill had already been introduced in the House to make the Naval Observatory responsible for distributing Washington time throughout the country, a time that would have to be used by railway companies in all their public schedules.

Allen took steps at once that would allow the railroad industry to participate in the process—via its own venue, the General Time Convention, which met every six months. Industry upheavals led the GTC to cancel its April 1882 meeting, so for months Allen could do little more than write editorials in the *Travelers' Official Guide*. Then, in October, at what some were calling the GTC's last meeting, he succeeded in getting its members to agree to meet the following April for the purpose of considering standard time for rail operations, a technical issue they had never considered before.[15] Among Allen's arguments was the near certainty of intervention by the federal government, particularly if the railroads themselves took no action. Undoubtedly he pointed to the joint resolution and to Fleming's proposal for a North American convention as evidence.

At the April 1883 meeting Allen convinced attendees to adopt for their operations a system of four local times separated by exact hours. In the months following he gained support from almost all the railway companies; he then reported his success at the October meeting of the General Time Convention. The industry agreed to make the change promptly.

On 18 November 1883, precisely at noon along the 75th meridian west of Greenwich, North American railroads switched to the new system, which they called Standard Railway Time. Immediately many American cities enacted ordinances making these same hour differences their civil times, thereby creating large regions—zones—having the same time throughout. Clocks were reset across the entire country, and more and more timepieces showed identical minutes and seconds; only the hours were different. A revolution in reckoning time uniformly was underway.[16]

Almost exactly two weeks after the railway companies shifted to Standard Railway Time, the secretary of state invited all countries that had diplomatic relations with the United States to send delegates to an international conference to be held in the City of Washington on 1 October 1884.[17] His timing of the American government's action was deliberately linked to the discussions at the Seventh General Conference of the International Geodetic Association, which had just met in Rome.

The importance of these discussions cannot be overestimated.[18] It is probable that without support from the specialists meeting there, the International Meridian Conference would not have taken place. Moreover, the deliberations at Rome are of keen interest in themselves, for they shed light on the issues discussed at the Washington conference.

As mentioned in the previous chapter, the Italian Geographical Society's circular letter to geographical and other scientific societies generated mixed results. Several respondents voiced opposition to a conference of government-accredited delegates. Among the reasons given were the difficulties inherent in agreeing on a common initial meridian and the long-standing preference held by many for the Greenwich meridian.

Among those favoring Greenwich were the members of the Geographical Society of Hamburg. The society's president, Gustav Heinrich Kirchenpauer, a member of the *Senat* of the free city-state of Hamburg and chief of the Department of Commerce and Navigation, wrote to the Permanent Commission of the International Geodetic Association asking for consideration of the subject of a common initial meridian and a decision regarding its use.[19]

The International Geodetic Association (Europäische Gradmessung) was an organization of member nations founded in 1862 when the king of Prussia decided to establish a scientific organization devoted to determining the figure (shape) of the earth and the various geophysical parameters associated with that shape to the highest levels of precision and accuracy. The king in turn was acting on an 1861 proposal from the geodesist and Prussian general Johann Jacob Baeyer to measure a central European arc, which meant that numerous astronomical observatories throughout the region would be linked, via existing and planned triangulation networks associated with the large-scale topographical surveying that was already being undertaken by various national governments.

Eighteen European states and countries agreed to participate in this proposed pan-European effort. At the IGA's first general conference in 1864, the technical experts there, besides creating an organizational structure, agreed on an extended research program and arranged to hold general meetings every three years. Though the IGA's final resolutions were strictly advisory in nature, member governments generally accepted them.[20]

Months before the request from Hamburg arrived, the Permanent Commission had agreed on the technical program for the next general conference. Consideration of an initial meridian did not fall within the agreed-on research agenda; indeed, the subject itself was of scant interest to geodesists. Nevertheless, IGA members had the right to propose subjects for inclusion at general conferences, so in deference to Kirchenpauer's late-November request, on 18 December 1882 the officers of the Permanent Commission addressed a letter to its seven members. They requested a formal agreement to consider a common initial meridian at the Rome conference, where a decision regarding its selection would be made.[21]

The tone of the letter was strongly in favor of placing this topic on the conference agenda. The officers noted that although the issue of an initial meridian did not involve geodesy directly, the subject was an important one, not only from the general view of civilization, but for progress in the geographical sciences—which the IGA certainly supported.

The IGA reminded fellow members that it was one of the era's leading international organizations, one through which governments and local civil authorities undertook research in fields that by their very nature crossed political borders. Accordingly, the IGA had an interest in arriving at a unification of longitudes, not only because of many concerns of the first order—navigation, maritime commerce, and the administration of telegraphs and railways—but also to facilitate the teaching of geography and for cartography, both of which suffered greatly from the diversity of base meridians.

The letter also noted the IGA's contributions to the determination of longitude differences by telegraph and that, as a result, the European continent was now covered by a reliable net of precise position values. The officers of the Permanent Commission expressed their confidence that certain progress would ensue if all longitudes were linked to the same initial meridian.

The letter also addressed the question of a universal time. According to its authors, European countries urgently needed a universal time for the internal operations of their telegraph and railway systems—a specialized time independent of local and national times.

Also believing that no progress could be made toward establishing a common initial meridian without the participation of Great Britain and the United States, the Permanent Commission's officers asked the other members if they would agree to have these countries' experts participate as special delegates. Further, members were asked to consult with the bureau responsible for publishing their country's astronomical and nautical almanacs to gain their technical views prior to the meeting. Finally, each member was asked to prepare a report that could serve as the basis for discussing issues, particularly scientific and technical ones.

Given this document, it was no surprise that all but one of the Permanent Commission's members agreed to have the subject placed on the agenda of the IGA's Seventh General Conference.[22] Thus, on 15 October 1883, thirty-two delegates from fourteen member states assembled in the hall of the Palazzo dei Conservatori on Rome's Capitoline Hill. The special delegates from Great Britain were the astronomer royal and director of the Royal Observatory at Greenwich, William H. M. Christie (1845–1922), and the geodesist Colonel A. R. Clarke, recently resigned from the Ordnance Survey. General R. D. Cutts, first assistant in the U.S. Coast and Geodetic Survey, represented the United States. In addition, the Permanent Commission invited the astronomers responsible for three of Europe's most important astronomical ephemerides and nautical almanacs: Wilhelm Foerster, from Berlin; Maurice Loewy, director of the *Connaissance des temps*; and Cecilio Pujazon, from Madrid. Twelve of the thirty-eight attendees were directors of national astronomical observatories, and the rest were geodesists, topographical surveyors, hydrographers, and specialists in other aspects of astronomy.[23] All in all, it was a most prominent gathering.

After some preliminaries, the designated *rapporteur*, Adolphe Hirsch (1830–1901; see Figure 5.2), one of the Permanent Commission's secretaries, director of the Neuchâtel Observatory in Switzerland, and the central figure in the sessions devoted to the subject, presented the "Report on the Unification of Longitudes by the Adoption of a Unique Initial Meridian, and on the Introduction of a Universal Time."

Both Hirsch and the president of the Permanent Commission, General Carlos Ibáñez of Spain, were longtime officers (secretary and president, respectively) of the International Committee for Weights and Measures (CIPM). This group of experts had been established by the Convention of the Metre to oversee the operations of the International Bureau of Weights and Measures (BIPM). Undoubtedly Hirsch had in mind the CIPM's responsibilities to member states who had signed the treaty bringing the BIPM into existence, and whose delegates now comprised the overarching diplomatic organization (General Conference on Weights and Measures, or CGPM). He declared at the outset that the IGA members at Rome had authority from their governments to discuss the issues, that their mission was official but not diplomatic, and that their efforts were scientific and preparatory ones. He acknowledged that whatever resolutions might be adopted in Rome were not obligations, but he looked

FIGURE 5.2

Adolphe Hirsch, director of the Neuchâtel Observatory, ca. 1880. The Swiss astronomer led the session on selecting an initial meridian at the IGA's Seventh General Conference, which met in Rome in 1883. President and Fellows of Harvard College; from HOLLIS #006131312.

forward to their consideration at some future diplomatic conference of the kind already proposed by the United States.

Having set the parameters, Hirsch turned to the first technical part of his presentation, "Unification of Longitudes by the Choice of a Unique Initial Meridian." Not surprisingly, he offered the same arguments in favor of a single meridian espoused by geographical scientists, astronomers, and meteorologists, as well as by those concerned with teaching geography in primary and secondary schools and in schools of navigation.

More interesting was Hirsch's enumeration of the possible disadvantages of change: the need to modify many astronomical ephemerides and nautical and astronomical almanacs, and to revise numerous collections of tabulated longitudes of geographical positions. Hirsch noted an even more serious disadvantage: the need to shift the longitude lines printed on the maps of many countries. This he sidestepped, however, by commenting that changing to a common initial meridian would not affect any charts or maps produced after the changeover.

Invoking macroeconomics, Hirsch reminded his audience how millions were being saved by the use of a decimal system for weights and measures and remarked that the savings accruing from meridian reform, while obviously not of the same magnitude, would nevertheless mean a real saving in time and work. Predictably, Hirsch concluded that the advantages of changing to a common initial meridian outweighed the disadvantages.

Deliberately going beyond the careful pronouncements of geographical congresses and other professional bodies, Hirsch declared that simply stating that unification would be useful and desirable would not satisfy their mandate, which was to arrive at a solution. Thus began the discussion of specific choices for an initial meridian. For many attendees, this would be the first time they had ever seriously compared the competing proposals.

Hirsch's first criterion was that the meridian should be determined at an astronomical observatory of the first order. Ocean navigation required calculations exact to one-half minute of arc (two seconds in time), equivalent to an uncertainty in position of one kilometer at the equator. Naturally, the accuracy desired—and achievable—in the geodetic and astronomical sciences was significantly higher: down to several hundredths of a second, or tens of meters. Consequently, it was not good enough to fix the world's initial meridian by an island (e.g., Ferro), a strait through which it might traverse, or even a mountain peak (Pico de Tenerife) or some monumental structure (the Great Pyramid at Giza). Moreover, only an observatory working at the state of the art could determine the stability of the instrument used to realize the initial meridian, in terms both of its mechanical and optical construction and of the site's underlying geology. Further, because any observatory's record of stability was a short one, Hirsch argued for the convenience of linking the point of departure of the world's terrestrial meridians directly to the ensemble of observations made by other observatories of the first order, each in turn linked via networks of precisely measured triangles (see Table 5.1).

Hirsch specifically dismissed de Beaumont's Bering Strait proposal: the geographer could not be serious, he said, in suggesting that an observatory be constructed in the middle of the strait (or on the isolated Prince of Wales promontory) and linked telegraphically to the neighboring continents for the sole purpose of realizing the initial meridian.

Hirsch then considered two specific arguments in favor of selecting an ocean meridian. The first was the assertion that the only way to reach global agreement was to choose a "neutral" meridian that passed through no country. In rebuttal, he reminded his audience that a country's adherence to the Convention of the Metre meant that its prior standards of length and weight had been discarded, with no apparent loss of national prestige. He asked the group of experts seated before him if any of them would seriously choose to build an observatory in the middle of an ocean—on one of the Fortunate Islands, perhaps, and maintain it with funds from all governments—merely to create an initial meridian that would not be English, French, German, or American.[24]

As to the other argument supporting an ocean meridian—that European countries might suffer from having to compute east and west longitudes (or their equivalent, plus and minus values)—Hirsch proposed that longitudes be computed in one direction only, starting from 0° at the initial meridian and increasing eastward around the globe to 360°.

TABLE 5.1

Initial (prime) meridians proposed, 1870–1875

Meridian	Degrees from Greenwich
Jerusalem	35°13'E
Great Pyramid, Giza	ca. 31°10'E
Cape Prince of Wales extended over the poles	11°55'E
Paris	2°20'E
Greenwich	—
Tenerife, Canary Islands	16°34'W
Ferro (Hierro)	17°40'W
60° from Alexandria	26°10'W
Greenwich + 2 hrs	30°00'W
Cape Prince of Wales	168°05'W
Bering Strait	169°00'W
Greenwich + 12 hrs	180°

Having reviewed the pros and cons of particular choices in detail, Hirsch concluded that, of the world's major meridians—Greenwich, Paris, Berlin, and Washington—none appeared superior to the others in any scientific aspect. Thus, Hirsch concluded, the decision turned solely on practical criteria: which meridian stood the best chance of being generally accepted, or at least adopted, by most "civilized" (Western) countries? Which choice would require the fewest changes in maps and charts, almanacs, manuals and handbooks, and geographical collections? His choice, of course, was Great Britain's Greenwich meridian.

Hirsch pointed out the vast area and population of the British Empire, the size of its naval and merchant fleets, and the fact that the Greenwich meridian was also used by the United States, Germany, Austria, and Italy. He estimated that 90 percent of the world's navigators calculated their longitude in terms of Greenwich.

Hirsch added that ephemerides based on the Greenwich meridian were also just as widely used and were produced in vastly greater numbers than those of the Paris-based *Connaissance des temps*. Moreover, those topographical surveys and, above all, the hydrographic ones laid out on the basis of the Greenwich meridian encompassed a terrestrial surface greater than the sum of all charts based on every other initial longitude. For all these reasons, to avoid the difficulties sure to be faced in changing it, he proposed that Greenwich be selected as the universal prime meridian.

Hirsch then turned to the second section, "Unification of Time." Establishing first the subject's direct link to longitude, he predicted that, when all ephemerides were calculated in terms of one and the same meridian, the time of that meridian would be the one employed throughout the fields of astronomy, geodesy, meteorology, and geophysics—thus creating a worldwide universal time.

However, he went on, there was a further issue to consider, namely, the need for a universal time in international communications: postal services and telegraph, railway, and steamship lines. Although he provided no concrete examples, Hirsch insisted that all those responsible for maintaining and expanding such public services were anxious to embrace a unique and universal time for their operations.

As a supporting argument, he noted the inconvenience caused when a country adopted a national time. Citing the Berlin astronomer and special delegate Wilhelm Foerster's recent article on the subject, Hirsch asserted that trying to employ a single (national) time throughout vast countries such as Austria-Hungary, Russia, or the United States was intolerable.

Hirsch opposed adopting standard (regional) times of the sort proposed by Sandford Fleming that divided the globe into twenty-four sectors. He also dismissed the system expounded by the Swedish astronomer Hugo Glydén: to create time changes at ten-minute intervals, producing 144 zones around the globe.[25] In Hirsch's view, the first scheme would lead to difficulties for any railway that had stations in different hour regions, while the latter was simply too complex to be workable for railways, telegraphs, or postal services. Moreover, Hirsch insisted, all the sciences would suffer if a regional system of multiple times were introduced.

His solution, one satisfying all needs, was to establish one single, universal time that would coexist with local time or a given country's national time.

Admitting that having two times was somewhat inconvenient and could certainly lead to errors, Hirsch proposed counting universal time from 0 hours to 24 hours, while local time should continue with the day subdivided into two groups of 12 hours, each as A.M. and P.M. Furthermore, the two different times would be used only by particular employees—train conductors, stationmasters, bureau chiefs, and the like—who could procure special watches with double dials. Most workers and ordinary citizens would not be affected and would not need to alter their long-held way of telling time, for local time was not being supplanted.

Turning to the final issue—where to begin the universal hour and universal day—Hirsch reminded his audience of the "unfortunate difference" between the civil day, beginning at midnight, and the astronomical day, which commenced the following noon.[26] He saw no useful reason for continuing this difference but admitted it was probably impossible to win agreement among astronomers to shift the start of the astronomical day from noon to midnight.[27] On the other hand, any attempt to convince the general populace to change their civil date to begin at noon was simply bound to fail.

To solve this dilemma, Hirsch proposed that universal time be regulated by the meridian 180° distant from Greenwich, that is, to start the universal day exactly at midnight along that meridian—which would of course be equivalent to noon at the Greenwich meridian. Doing this would make the universal day and the astronomical day coincide and consequently minimize changes to ephemerides and almanacs. Moreover, aligning the universal day in this manner would retain ocean navigators' near-universal practice of changing their date when they crossed the 180th meridian.

Having finished the technical portions of his presentation, Hirsch, in the name of the Permanent Commission, closed by offering seven resolutions for the attendees to consider.[28] General Ibáñez, as president of the Permanent Commission, then appointed a seven-member Special

Commission to examine Hirsch's report and resolutions, discuss the various issues, and report back. This latter group, along with other interested members, met three times, debated the issues, voted to modify the wording of several proposed resolutions, added two more,[29] and finally summarized their findings at a general session of the conference. Here the recommended resolutions (now nine in number) were discussed briefly, voted upon separately, and then approved as a group. They became "The Resolutions of the International Geodetic Association, Concerning the Unification of Longitude and of Time."[30]

Throughout the ensuing deliberations of the Special Commission, it was clear that almost all attendees favored unifying longitudes, and that the vast majority favored Greenwich. Because so many European countries already used the Greenwich meridian, particularly on hydrographic charts, the economic burden associated with change would fall most heavily on France (and to a lesser extent, on Spain and Portugal). French delegates expressed their concerns in this regard and voiced their opposition to Greenwich. Consequently, two adjustments were approved to accommodate the French and achieve something near unanimity among the specialists.

First, all references in the resolutions to "topographic bureaus" were removed. This meant that their particular products—large-scale maps with base longitudes in terms of national observatories—would not be covered by the IGA resolutions.[31]

Second, a nontechnical resolution was introduced (Resolution No. VIII). This expressed the hope that if the civilized world agreed to adopt Greenwich as the unifying meridian for longitude and time, Great Britain would see fit to advance the unification of weights and measures by signing the Convention of the Metre. Although the French hydrographer Adrien Germain had earlier scorned such a quid pro quo as demeaning to science, other views, undoubtedly those of the CIPM officials Hirsch and Ibáñez, now prevailed.[32]

Diplomatic considerations were at the core of the ninth and last resolution. As we have seen, during the previous October, the American government had inquired as to the desirability of holding an international conference focused on a common zero meridian for longitude and time. Many countries had responded positively; however, several European governments had not done so, some preferring to first examine the views and recommendations of the specialists who were meeting in Rome.[33] The ninth resolution, worded by Hirsch, informed the European member states of the Rome conference's results and expressed the view that a conference focused on steps to officially approve the conference resolutions should be held as soon as possible.

In the last moments of the final session, one French delegate rose to oppose the resolution. It was unnecessary, he insisted, in light of the actions already taken by the United States. Immediately Hirsch added a clause specifically linking the American actions to the future diplomatic conference. As revised, the resolution (No. IX) was approved unanimously.[34] A nice diplomatic maneuver, indeed![35]

The deliberations at the Seventh General Conference of the International Geodetic Association were now over. The efficiency of the IGA process certainly seems impressive, particularly when one remembers that unification of longitude and time represented only one of eight areas scheduled for the week of technical sessions. Equally impressive was the quality of the analyses displayed in discussions of the various issues. Indeed, after the Rome conference the chance that any initial meridian other than Greenwich would be adopted was virtually nil.

Following decades of discussions and suggestions in a variety of forums, in Rome specialists had at last overwhelmingly agreed on an initial meridian line and defined it as passing through the point midway between the pillars of the meridian instrument at the Royal Observatory at Greenwich. For the sciences everywhere and for transportation and communication companies, these scientists had also established a new construct, a universal time linked to a cosmopolitan day—a two-time system in which the technically based time did not interfere with public-time practices. Although the specialists meeting in Rome remained unaware of the imminent changes in North America's public timekeeping, which would nullify some of their considerations, their accomplishments were truly magnificent.[36]

CHAPTER 6

Washington and London and Beyond

[A]n invitation . . . to meet delegates from the United States and
other nations in an International Conference . . . for the purpose
of discussing and if possible, fixing on a meridian proper to be
employed as a common zero of longitude and standard of time-
reckoning throughout the globe.
—Secretary of State F. T. Frelinghuysen, 1 December 1883

The ink had scarcely dried on the official invitations to the International
Meridian Conference when criticism of the process arose.[1] Earlier the
editors of *Science* had supported querying countries first in order to
judge the level of support among them. Now they wrote that "arrangements for the meeting of this conference have been delayed far beyond
anything customary, even for diplomacy," and wondered why so late a
date for holding it had been chosen. The editors of *Nature* echoed their
remarks.[2]

One wonders why such criticisms were raised, since the lack of any
urgency regarding meridian and time issues was by this time well established (Fleming's remarks and those by the IGA notwithstanding).
Moreover, the October 1884 date chosen by the American government
reflected two significant considerations. First, Washington's oppressive
summers made it impractical even to consider convening an international
meeting in July and August. Second, no funds for the conference had
been appropriated by Congress, and shifting current-year monies to support it was simply impossible. Certainly the American editor-critics were
well aware that the federal government's appropriations process followed
a constitutionally defined and carefully choreographed path, one that
had begun in the House of Representatives. In the case of the bill for
civil expenses of the government for the fiscal year ending 30 June 1885,
nothing could take place until the 48th Congress convened, which oc-

curred on 3 December 1883—two days after the secretary of state issued his circular invitation.

President Chester A. Arthur's request for an appropriation of $10,000 for expenses of the October conference came in mid-April; final legislation was approved on 7 July 1884, a few days after the start of the 1885 fiscal year. After negotiations between the two houses of Congress, $5,000 was approved (around $90,000 in today's dollars); the delegates were directed to serve without compensation.[3] Included in the act was language that authorized the President "to appoint two delegates . . . in addition to the number [three] authorized by the act" approved in August 1882. Its inclusion may have stemmed from the Senate's earlier desire to have five American delegates.[4]

Delegate appointments were also criticized by the *Science* editors.[5] The premise of their remarks was that the United States must be represented by the "highest order of talent," in particular "men of the highest scientific authority" in the matters coming before the conference. Without such qualifications, proposals made by the American delegates would carry little weight with those sent by other countries, who presumably would be experts. After voicing the hope that this country's delegates would be able to converse in French and German, the editors turned to the qualifications of the two known selections: F. A. P. Barnard, president of Columbia College, and Commander William T. Sampson, assistant to the superintendent of the Naval Observatory.

Barnard, of course, was an ideal choice. Also president of the AMS, he had maneuvered to make the meridian conference a reality. And he moved easily in international venues. While the editors had no objections to Barnard on technical grounds, they opined that his "personal disability of extreme deafness ought . . . to have excluded him" from consideration, for this handicap would "practically prevent him from taking a leading and representative part" in conference deliberations.

Although the editors viewed Sampson as capable, he was "practically unknown in science, outside of a limited circle in the United States." Rather unfairly, they wrote that "aside from his being present on duty at the naval observatory there is very little reason why he should have been selected for this responsible scientific appointment, rather than any other line officer of the navy."

Disappointed that Julius E. Hilgard, superintendent of the USC&GS, had not even been considered, the editors urged that the secretary of war, Robert Lincoln, who was responsible for proposing a candidate, go outside his department and recommend some other highly regarded geodetic authority. He eventually chose Cleveland Abbe, chief scientist of the U.S. Signal Office.

After criticizing the secretary of the navy, William E. Chandler, for passing over Simon Newcomb in favor of Sampson, the editors proposed that the president of the National Academy of Sciences take advantage of the academy's charter requiring it to advise the government in scientific matters when called on and make recommendations to Arthur to strengthen the American contingent. The *Science* editors, probably already aware that the House-approved bill increased the authorized number of delegates, added that immediate legislation was needed to effect their proposal—a hopeless endeavor so late in the process, as anyone at all familiar with government affairs would have known.

At best, the editors were naive regarding delegate qualifications. It is well understood that technical prowess cannot be the only criterion for appointments to a government-sponsored body, especially one having significant nontechnical aspects. In this case, excellent technical guidance was available via the deliberations of the IGA at Rome. The appointments of Barnard, Sampson, and Abbe were more than sufficient.

The gratuitous advice from *Science* was ignored. A few days after the delegate expansion had been enacted into law, Arthur selected the well-connected and highly regarded amateur astronomer Lewis M. Rutherfurd of New York City, whose early career was in the law, and the railway editor William F. Allen of New York City and New Jersey, the prime mover in North America's adoption of Greenwich-linked Standard Railway Time. Both added valuable experience to the delegate pool.

Eight days before the conference was to begin, Barnard resigned. This was a severe blow to the government; the secretary of state, Frederick T. Frelinghuysen, had given him the title "Chairman of the Board of Delegates on the part of the United States to the International Time Meridian Congress" and expected the educator to take charge of the myriad arrangements associated with sponsoring and conducting an international meeting—a nearly impossible assignment for anyone who did not reside in Washington. In his resignation letter to the president, Barnard pleaded that college business required his personal attention, making it impossible for him to be away from New York City during the next three or four weeks.[6]

Barnard's replacement was retired Rear Admiral C. R. P. Rodgers (1819–1892), the former superintendent of the U.S. Naval Academy and scion of a celebrated navy family. Rodgers was eventually elected meridian conference president.

In what appears to be an example of less than stellar planning by the national government, Frelinghuysen then announced that the very first meeting of the American delegation as a group would take place on 29 September, a scant two days before the start of the meridian conference.

In addition, the Americans' inexplicable failure to provide a French-language stenographer for an international meeting right from the very start was to lead to a week's delay during the proceedings.[7] Perhaps some of this seeming ineptness stemmed from the involvement of two government departments, for the State Department had been forwarding the diplomatic correspondence it received to the Navy Department as early as mid-July. Further, the meridian conference would have been a minor event for State, which had not introduced the initial legislation.

The clause allowing up to two additional delegates applied to all participating countries. It proved a boon to Great Britain, which apparently was under pressure to include Sandford Fleming in its official party and eventually did so.[8] Its three primary appointments were all distinguished authorities: Captain Sir F. J. O. Evans (1815–1885), who had just resigned as hydrographer of the Royal Navy; the astronomer John C. Adams (1819–1892), director of the Cambridge Observatory; and the geographer and meteorologist Lieutenant General Richard Strachey (1817–1908), a member of the Council of India. For Great Britain, the IGA resolutions regarding the adoption of the Greenwich meridian for longitude and universal time certainly made its delegates' tasks relatively straightforward and pleasant ones.

France, on the other hand, was in a most difficult position. Given the overwhelming sentiment already shown for adopting Greenwich, it was easy to predict that a majority of delegates to the International Meridian Conference would follow suit; indeed, some governments saw this Washington conference as simply a ratification of the recommendations made by the government-sponsored technical experts at Rome. Accordingly, soon after the formal invitation from the United States arrived, the minister of public instruction and fine arts contacted France's Academy of Sciences, which had been considering the issues ever since the United States first raised the possibility of convening a conference.

The academy's Sections on Astronomy and on Geography and Navigation responded with a joint report proposing that it serve as the basis for the instructions to the French delegates.[9] Following this, the minister of public instruction and fine arts established a commission of nearly two dozen experts representing all interested sciences and services—railroads, telegraph, maritime interests, and commerce—along with the director of the Ministry for Foreign Affairs. They were charged with discussing and recommending what positions the French delegates should take to Washington. The hydrographer Édouard Caspari prepared the commission's eighteen-page report, which contained four resolutions and supporting arguments.[10]

By far the most important resolution concerned the attributes of a

common initial meridian, which the French termed the *méridien international*. It was to be both international in character and neutral, a meridian that did not cut through either the European or American continents.
(A meridian passing through the Bering Strait met these requirements,
and the ancient one through the island of Ferro came close to the French
ideal.)

A complementary resolution expressed a desire for nations to incorporate the new meridian in as many applications as possible, subject to the
particular interests of each country. It was undoubtedly a hedge to allow
selective noncompliance.

Insofar as a universal time was concerned, commission members
were not much interested in the proposed reform, seeing few advantages
from it. However, they did not oppose its adoption, so long as the same
condition of neutrality was upheld. Consul-General Albert Lefaivre and
Jules Janssen (1824–1907), director of the Meudon Observatory, came to
Washington with the set of resolutions as their instructions.[11]

Two other European countries, Russia and Spain, included technical
experts in their delegations, while Austria-Hungary, Germany, Italy, the
Netherlands, and Sweden were represented solely by diplomats. Belgium,
Denmark, Portugal, and Serbia did not attend. Adolphe Hirsch, who had
guided the deliberations at the Rome conference so brilliantly, was one
of two expected to attend on behalf of Switzerland and would certainly
have been selected as one of the International Meridian Conference's
secretaries. However, he never arrived. Subsequently, three other experts—Luis Cruls, director of Brazil's Imperial Observatory, Jules Janssen of France, and Great Britain's Strachey—were nominated and, after
a unanimous vote, became the conference secretaries.[12]

Twenty-five nations were represented at the International Meridian
Conference, which opened at noon on Wednesday, 1 October 1884. All
sessions were held in the Diplomatic Room at the State, War and Navy
Building.[13] Like the Rome conference, delegates at the meridian conference had no authority to commit their nations to any resolutions. However, in contrast to the process at the IGA conference in Rome, where
voting had been by individuals, voting at the International Meridian
Conference was by countries. Delegates had received instructions from
their governments regarding specific positions to take in Washington.

After the preliminaries, the first days of the meridian conference were
dominated by France. The French delegates insisted on what they argued
was a logical development for the process. Consequently, specifics regarding the choice of an initial meridian were held in abeyance while
a general resolution—that the delegates viewed it desirable "to adopt a
single prime meridian for all nations, in place of the multiplicity of initial

meridians which now exist"—was introduced, discussed briefly, and accepted unanimously.[14] This general resolution was essentially identical to the initial one approved at the Rome conference.

Then the French made formal their key instruction, proposing that the world's initial meridian be an absolutely neutral one, chosen in a manner to bring all possible advantages for science and the international community, and not pass across any great continent.[15] As was obvious to all, this motion ran completely counter to the Rome conference discussions by excluding all meridians linked to observatories of the first order: those of Berlin, Paris, Greenwich, and Washington.

Two days were spent discussing the motion, the French delegates presenting arguments in its favor and the British and American delegates asking for details about how an absolutely neutral location could be established. The astronomer Simon Newcomb, one of the distinguished invited guests attending the conference, commented that from a practical standpoint, no absolutely neutral prime meridian could be envisioned, for any choice would have to be referred to some national observatory. Others ignored any philosophical implications inherent in the motion, articulating instead the advantages in convenience and economy of selecting Greenwich. Sandford Fleming presented his now-familiar tabulation of ship tonnage versus a country's initial meridian to show that Greenwich was already the most widely used, then quoted Otto Struve's 1880 report in support of the meridian 180° from Greenwich. Fleming asserted that by adopting the Greenwich anti-meridian, those concerns regarding the initial Greenwich meridian's national character would be greatly reduced and the benefits of a Greenwich meridian for almanacs and charts retained. His "compromise" fell on deaf ears.

The French delegates made it clear that defeat of their motion would lead them to abstain from voting on any meridian-related resolution. But no compromise was possible. A vote was taken, and the motion was defeated by a margin of twenty-one to three, Brazil and San Domingo (Haiti) voting with France.[16]

Attention then shifted to the choice of a common meridian. Earlier, the American delegate Lewis Rutherfurd had introduced a resolution proposing the adoption of the Greenwich meridian for reckoning longitude, a proposal that had been withdrawn temporarily to allow the French to present their motion. Presumably acting as spokesman for his government, he now reintroduced it. He also linked it to a group of related proposals in order to give the other delegates a common framework for subsequent discussions.[17] The proposed resolutions were almost identical in content to those adopted at Rome, but with significant technical changes in two of them (see Table 6.1).

TABLE 6.1A

International Geodetic Association resolutions, 1883

I.	The unification of longitude and of time is desirable. . . . recommended to the Governments of all States interested . . . that hereafter one and the same system of longitudes shall be employed in all the institutes and geodetic bureaus . . . [with exceptions].
III.	The Conference proposes to the Governments to select for the initial meridian that of Greenwich . . . [which] fulfills . . . all the conditions demanded by science . . . and . . . it presents the greatest probability of being generally accepted.
IV.	It is advisable to count all longitudes, starting from the meridian of Greenwich, in the direction from west to east only.
V.	The Conference recognizes for certain scientific wants and for the internal service in the chief administrations of routes of communication . . . the utility of adopting a universal time, along with local or national time. . . .
VI.	The Conference recommends, as the point of departure of universal time and of cosmopolitan date, the mean noon of Greenwich, which coincides . . . with the commencement of the civil day, under the meridian situated 12 hours or 180 degrees from Greenwich. It is agreed to count the universal time from 0h to 24h.
VII.	[T]he States which . . . find it necessary to change their meridians, should introduce the new system of longitudes and of hours as soon as possible.
II.	Notwithstanding the great advantages . . . [of] decimal division . . . in . . . co-ordinates . . . and in the corresponding time expressions . . . it is proper to pass it by. . . .
VIII.	The Conference hopes that if the entire world . . . [accepts] the meridian of Greenwich . . . Great Britain will find in this fact . . . a new step in favor of the unification of weights and measures, by acceding to the Convention du Mètre of the 20th [of] May, 1875.
IX.	These resolutions will be brought to the knowledge of Governments and recommended to their favorable consideration. . . .

SOURCE: Philosophical Society, "Resolutions." The sequence of the resolutions has been altered to facilitate comparisons.

During the relatively brief discussion of the first resolution, two issues arose. First, Sandford Fleming tried to modify the description of the Greenwich meridian by aligning it within the great circle that included its anti-meridian—doing so in anticipation of, and to allow for, his great plan to count longitude from the latter and also define universal time in terms of it. Fleming presented his view as a substitute amendment. Immediately John C. Adams indicated that he and the other members of the

TABLE 6.1B

International Meridian Conference resolutions, 1884

I.	That it is the opinion of this Congress [*sic*] that it is desirable to adopt a single prime meridian for all nations, in place of the multiplicity of initial meridians which now exist.
II.	That the Conference proposes to the Governments here represented the adoption of the meridian passing through the center of the transit instrument at the Observatory of Greenwich as the initial meridian for longitude.
III.	That from this meridian longitudes shall be counted in two directions up to 180 degrees, east longitude being plus and west longitude minus.
IV.	That the Conference proposes the adoption of a universal day for all purposes for which it may be found convenient, and which shall not interfere with the use of local or other standard time where desirable.
V.	That this universal day is to be a mean solar day; is to begin for all the world at the moment of mean midnight of the initial meridian, coinciding with the beginning of the civil day and date of that meridian; and is to be counted from zero up to twenty-four hours.
VI.	That the Conference expresses the hope that as soon as may be practicable the astronomical and nautical days will be arranged everywhere to begin at mean midnight.
VII.	That the Conference expresses the hope that the technical studies designed to regulate and extend the application of the decimal system to the division of angular space and time shall be resumed, so as to permit the extension of this application to all cases in which it presents real advantages.
	That a copy of the resolutions passed by the Conference shall be communicated to the Government of the United States of America. . . .

SOURCE: U.S. President, *Message . . . Relative to International Meridian Conference*, 111–112.

delegation would vote against it; that is, the government of Great Britain was opposed. When put to a vote, Fleming's amendment lost.[18]

Then in an echo of the Rome conference, the diplomat-delegate from Spain announced that his country would vote in favor of adopting the Greenwich meridian, expressing at the same time his government's hope that Great Britain would accept the metric system of weights and measures; he indicated that Italy held a similar view. When questioned, the

Spanish delegate stated that his vote was not conditional on Britain's acceptance; however, he had been instructed to voice that hope.[19]

Strachey responded to the Spanish delegate. As a result of the hope expressed by the IGA's Rome resolution, the government of Great Britain did examine its current position carefully and decided to adhere to the Convention of the Metre. All arrangements were complete, or nearly so, he reported. (Official adherence came on 11 September 1884.)[20]

Of course the British decision did nothing to advance adoption of the Greenwich meridian by France. Its astronomer-delegate continued to complain that his country had before it a resolution not linked primarily to science, and one that required a sacrifice of French tradition as well as imposing a financial burden: costs of revisions to its maritime products. France, he said, could not and would not concur in this resolution.

At this point the physicist Sir William Thomson, another distinguished non-delegate attendee, was invited to speak. He attempted to mollify the French, remarking that "England is making a sacrifice in not adopting the metrical system," a view not germane to the issue at hand. And he, too, saw the adoption of Greenwich as based on general convenience, not science.

Earlier the French delegate Janssen had cited a hydrographic work that reflected on the cost that would accrue to France if Greenwich were selected. Its 2,600 hydrographic-chart copper plates, over 600 of its sailing directions, and sections of the *Connaissance des temps* would have to be altered. Partly to counter this complaint, but also to document the importance of the Greenwich meridian among the maritime nations of the world, the delegate Sir F. J. O. Evans showed that the annual sales of 100,000 Admiralty charts included purchases by a host of countries, France included. He noted that 2,900 copper plates were in constant use, with 60 new ones being added each year; added that commercial firms sold large numbers of Greenwich-linked charts printed from some 930 of their own plates; stated that the Admiralty published fifty-one volumes of sailing directions; and gave figures showing that annual sales of the *Nautical Almanac* exceeded 15,000.

Evans was immediately followed by the United States delegate and railway editor W. F. Allen, who argued that the Greenwich meridian had come to be of great importance to American commerce. He urged that no switch in initial meridians be made, for doing so would require his industry's Standard Railway Time meridians to be redefined in terms of fractional hours.[21]

As anticipated, the vote in favor of adopting Greenwich as the world's initial meridian for longitude was overwhelming: twenty-one in favor, San Domingo opposed, with Brazil and France abstaining. (Subsequently the delegate from [El] Salvador, who was absent during the official polling of delegates, added his country's vote to those favoring Greenwich.)

The conference delegates next turned to the second longitude-related proposed resolution (No. III). Instead of counting longitudes in one direction only from the initial meridian, as urged in the resolution adopted at Rome, this proposal was to count longitudes in both directions from Greenwich, with east longitude being plus and west being minus.[22]

The reason for the proposed change was that opposition had surfaced subsequent to the Rome conference. While the American delegate to the IGA's conference was summarizing the actions that had just been taken there, Superintendent Hilgard of the USC&GS announced that "[i]t will be impossible . . . to agree to the rule which counts all longitudes from west to east."[23] And so the American position, as introduced by Rutherfurd, came before the delegates in Washington.

The United States was not alone in its view. C. de Struve, speaking for Russia's delegation of diplomats and specialists, pronounced it the more practical approach, as well as one that would continue the longtime use of specifying longitudes in both directions. Moreover, the discontinuity in numbering on reaching 180° would take place at the same location where mariners historically changed the day in their reckoning.[24]

During the session an amendment conforming to that approved at Rome was introduced: to count longitudes in one direction, from west to east. This gave rise to another amendment: to have the counting go not from west to east, but from east to west. "A matter of detail," several delegates suggested, and the debate dragged on. After an adjournment to allow delegates to discuss the issues informally, both amendments were withdrawn. A majority of delegates argued that no reason, scientific or otherwise, existed to change mariners' common usage, which was to count longitudes in both directions from the initial meridian to the anti-meridian.

Just before the vote on the resolution, Sandford Fleming, still promoting his system of universal time linked to the anti-meridian of Greenwich, spent considerable effort outlining his specific proposals. Once again the British delegation's majority opposed his ideas.

Counting longitudes in both directions, the position of both the United States and Great Britain, was adopted: fourteen in favor, five opposed, with six countries abstaining. The smaller size of this majority vote reflected both the fact that many delegates had been instructed to favor the resolutions adopted at the Rome conference and the influence of non-European delegates who had come with no official position. Indeed, all European nations except Great Britain and Russia abstained or voted against the resolution.

The delegates now addressed the question of the universal day, a subject that had already colored sessions associated with adopting a common initial meridian. As expected, a general resolution (No. IV) confirming the need for a universal day, "for all purposes for which it may be found convenient," met little opposition.[25] However, the American delegate W. F. Allen introduced an amendment and then launched into a detailed summary of the previous year's adoption of hour sections by North American railroads, going on to show their extension to the rest of the world. Probably this action was simply a ploy to place information in the official record, for the railway editor withdrew his amendment (which was not germane to the discussion) as he concluded his remarks. After an unimportant Italian amendment was defeated, the general resolution was adopted, the vote being twenty-three in favor, with two abstentions.

The next resolution (No. V) presented delegates with another proposal different from the resolution adopted at Rome. Here it was changing the beginning of the universal day from noon at the initial meridian to midnight to conform with Greenwich civil time.[26] Because of this most significant change, much time was spent on amendments; and as we shall see, the final vote clearly reflected delegates' instructions from their governments.

Early on, the senior Spanish delegate, who had been instructed to support both the concept of a universal meridian and the adoption of Greenwich specifically, announced that he had no authority from his government to discuss the proposed resolution, and further, because this was a scientific question, he felt he was not competent to judge it. He urged that voting be put off to allow time for reflection.

At the next session another Spanish delegate, in what can only be described as a confused, lengthy presentation, proposed an amendment: that the meridian conference abstain from adopting any meridian for time. Since a resolution calling for the adoption of a universal time had already been adopted, this proposed amendment was defeated, apparently by voice vote.

Another amendment, by the Spanish delegation's hydrographic specialist, Juan Pastorin, would have located the prime meridian for cosmic (i.e., universal) time at the Greenwich anti-meridian, and divided the surface of the globe into twenty-four-hour bands, with the cosmic hours running from zero to twenty-four in the direction counter to the earth's rotation—that is, east to west from Greenwich's anti-meridian. His amendment, of course, was the same proposal made by Sandford Fleming.[27] Discussed during a brief recess, it, too, was defeated by a voice vote.

The last amendment considered, introduced by the delegate from Sweden, was the resolution adopted at the Rome conference: that the initial point of the universal hour and cosmic day be the moment of mean midnight at the anti-meridian of Greenwich—the start of the civil day there. Once again it was pointed out that its adoption would mean that during the twenty-four-hour civil day of France, England, and all other European countries, their morning hours would have one universal date, while their afternoon hours would have another, for under the amendment, the universal day would begin at *noon* at the Greenwich meridian.

Delegates were reminded that one reason this resolution had been adopted at Rome was to ensure that no changes need be made in the various ephemerides and almanacs, whose basis was the astronomical day. But the meridian conference attendees argued that the choice should be based on what would cause the least confusion and inconvenience, not to astronomers, but to the world at large. Since it was anticipated that the communications and transportation industries would use the universal day—at least in their internal arrangements—making it consistent with the Greenwich civil day seemed the logical choice. As another distinguished non-delegate, the astronomer W. Valentiner, pointed out, "The object is to introduce uniformity in time-reckoning in the astronomical and the civil world . . . [so] it is the astronomer only that must give way." And the astronomer-delegate John C. Adams cited numerous astronomical tables that began the day at midnight to document that beginning the day at noon was not universal even among his colleagues.[28] When put to a vote, the amendment—the Rome conference resolution—lost, with six in favor, fourteen opposed, and four countries abstaining.

Now the resolution to have the universal day coincide with the beginning of the civil day and date at the Greenwich meridian was before the delegates. Sandford Fleming, whose own proposals had all been rejected in previous sessions, now spoke in favor of adopting this one. The vote was fifteen countries in favor, two opposed (Austria-Hungary and Spain), and seven abstaining, a vote later changed to fourteen ayes, three nays, and seven abstaining. Once again, among the European countries only Great Britain and Russia voted for the resolution.[29]

At this moment it was late in the afternoon of the meridian conference's sixth session. With the vote just concluded, all conference goals were met: a common initial meridian for longitude, and a standard for time reckoning. With many formal housekeeping details still before them at what was the start of a fourth week in Washington, delegates must have been anxious to wrap up their efforts as swiftly as possible.

Into this environment Rutherfurd offered the last of his proposals: a hope that one day the astronomical and nautical days would begin at midnight, thereby conforming to the civil day. A trivial adjustment to the words was made, and then, apparently with no discussion at all, the resolution (No. VI) was "carried without division." Not one of the thirty-nine delegates there recalled the admonition of the Spanish delegate that "we do not insert any hopes in our protocols."[30] As we shall see in the next chapter, the adoption of this resolution proved to be a major blunder.

Before adjourning for the day, the delegates also extended a courtesy to the French delegation. They overrode a "not-germane" ruling by the meridian conference's president and adopted a resolution (No. VII) expressing the hope that studies designed to extend the decimal system to angles and time would be resumed. This topic, discussed at the Rome conference and sidelined there, was part of the French government's instructions to its delegates to the International Meridian Conference.[31]

Also in the closing moments of the session, the delegates were confronted by a new subject: that successive meridians around the globe be used for civil timekeeping and linked to the prime meridian. This proposal was broad enough to include both the one-hour-difference and the ten-minute-difference proposals. After agreeing to insert a report on the latter written by the astronomer Hugo Gyldén and offered by the delegate from Sweden, the delegates adjourned. At the next session, however, the resolution was withdrawn. Thus the concept of time zones was never part of the International Meridian Conference's deliberations.

The rest of this session and the final session on 1 November 1884 were devoted to housekeeping tasks: approval of the French- and English-language protocols, adoption of a final act summarizing the seven resolutions, and passage of a resolution directing that a copy of them be "communicated to the Government of the United States of America, at whose instance and within whose territory the Conference has been convened." Finally, thirty-one days after the inaugural session, the International Meridian Conference was over.[32]

As already noted, delegates to the meridian conference in Washington were not authorized to commit their countries to the resolutions; another assembly of representatives having full powers to do so was required. The first steps leading to such a meeting were taken on 1 December via a brief summary in the *Annual Message of the President*, followed by a *Message from the President of the United States*, which contained the text of the conference proceedings. Then early in February of 1885, less than a month before the end of second session of the 48th Congress, Frelinghuysen sent a letter to the chairman of the Senate Committee on Foreign Relations stating that President Arthur was of the opinion

that he had done all that is necessary to bring the matter again within the jurisdiction of Congress (where the project originated), and that it is open to that body to signify its wish as to whether the conclusions reached by that Conference shall be brought by this Government formally to the notice of the other Governments, with an invitation to adopt them for universal use by means of a general international convention to that end.[33]

A draft of a joint resolution accompanied the secretary of state's letter.

On 9 February the Senate passed a concurrent resolution authorizing the president to communicate with all countries regarding the resolutions adopted at the International Meridian Conference and to invite their accession.[34] And, during that same week, Secretary of the Navy Chandler formally proposed to the Senate that, in light of the meridian conference's resolution recommending a common prime meridian, the U.S. statute requiring two American meridians be repealed. In his letter he included the draft of a replacement: "The meridian of Greenwich shall be adopted for all nautical and astronomical purposes."[35]

A few days later the navy secretary released to the Senate twelve pages of communications associated with what had already become an extremely controversial issue among the country's astronomers: altering the start of the astronomical day. While indicating that he was submitting the material in connection with the meridian conference's recommendation for a common prime meridian, Chandler gave no indication of the executive branch's position on this subject.[36]

One suspects that the flurry of activity within the executive branch in February 1885 was no more than housecleaning—finishing outstanding but unresolved tasks. Indeed, with a controversy now swirling around one of the meridian conference's resolutions, progress was no longer a simple affair—particularly at such a late date. Although a Senate committee report on the concurrent resolution—to inform other governments of the meridian conference's resolutions—had been prepared, much more was now required. Nothing was forthcoming. The second session of the Forty-eighth Congress ended without the House taking any action at all.

On 4 March 1885, Grover Cleveland, a Democrat, succeeded the Republican Chester Arthur as president of the United States. Without any doubt the new cabinet had no interest in championing legislation associated with the meridian conference's resolutions. Indeed, now that one facet of those resolutions had aroused controversy, one would scarcely expect any movement at all by the executive branch—certainly not until the cabinet secretaries were familiar with what they must have viewed as a complex technical matter.

With progress halted in America, the push for a common meridian for longitude and timekeeping switched to Great Britain. Undoubtedly its specialists were quite satisfied with the meridian conference's outcome. Unlike the situation for France, adopting Greenwich required no costly changes in their country's official maps, hydrographic charts, or sailing directions.

Soon after the official protocols reached London, they were sent to the lords of the Committee of Council on Education, under whose authority the Science and Art Department fell. Early in 1885 they appointed a senior-level technical committee to advise them on what actions to take.[37] This committee's initial advice led to consultations with various government departments, with learned societies, and with telegraph companies; all indicated unanimous approval of the meridian conference's first five resolutions. The committee then advised the lords of the Committee of Council on Education that no action on the government's part was required except on Resolution VI, conforming the astronomical day to the civil day. As we shall see, the British government's subsequent focus on this particular resolution merely compounded the blunder made at the International Meridian Conference itself.[38]

Not surprisingly, no progress was taking place elsewhere in Europe. Indeed, throughout the 1880s nothing happened anywhere—with the exception of Japan. On his return home, Japan's delegate, D. Kikuchi, dean of the science department of the Imperial University of Tokyo, reported favorably on the meridian conference's resolutions. A subsequent government committee also reported favorably on them, and on 12 July 1886 an Imperial Ordinance was signed. This established Greenwich as the country's first meridian for calculations of longitude and standard time, with the latter to be introduced on 1 January 1888 and to be nine hours in advance of Greenwich.[39] This action is judged to be the only tangible result of the International Meridian Conference.

In 1888, when Cleveland was nearing the end of his term, one last attempt was made by the United States to have the meridian conference's resolutions formally accepted by the nations of the world. In a communication to Secretary of State T. F. Bayard, Rear Admiral C. R. P. Rodgers, who had been president of the conference, summarized the resolutions and the subsequent lack of progress during the 49th Congress. Rodgers urged Congress to consider the matter again, approve the resolutions of the meridian conference, and further, authorize the president to invite the "powers with whom this country has diplomatic relations to accede" to them. Endorsed by Bayard and by Cleveland in a 9 January 1888 *Message from the President of the United States*, this communication came to the 50th Congress early in its first session. Referred to the House Committee on Foreign Affairs, it promptly disappeared from sight.[40]

More than three years had passed. The seven resolutions agreed to by the delegates to the International Meridian Conference had not gained formal acceptance even from the American government, which had convened this diplomatic effort. No process had been established that would lead to their adoption by the nations of the world; none was likely. The International Meridian Conference had clearly failed.

That others were aware of the United States' failure to resolve the prime meridian and universal time issues can be seen in renewed efforts mounted by advocates of non-Greenwich meridians. In 1888, in a memorial published in a Swiss geographical journal, de Beaumont traced the history of uniformity attempts since 1871. He next presented a personal and obviously false view: that the unrepresentative nature of both the IGA meeting at Rome and the International Meridian Conference made it impossible for their resolutions to be accepted. Not surprisingly, he continued by extolling the advantages of the Greenwich-plus-180° antimeridian, which he now called the "Bering Strait meridian" and linked to universal time.

At the Fourth International Geographical Congress meeting in August 1889 in Paris, R. P. Tondini de Quarenghi, an Italian missionary representing the Royal Academy of Sciences of Bologna, proposed to the congress's group I, Mathematical Geography, the meridian of Jerusalem as a neutral meridian and its use for the hours of universal time. After group I delegates declined to pass on de Quarenghi's and other proposals to the general session of the congress, the Bologna academy presented it via letter to the British Association for the Advancement of Science. Meeting in Newcastle-upon-Tyne, the BAAS committee charged with studying the subject wrote that "the question of a universal prime meridian is one that cannot usefully be considered . . . at the present time."[41]

Despite these rebuffs, the effort to make Jerusalem the first meridian for calculating universal time was not over. In July 1890 Italy's ambassador to the Court of St. James invited British government representatives to attend a congress in Rome under the aegis of his government to consider adopting an initial meridian—Jerusalem—to provide a uniform time all over the world. All the countries represented at the Washington conference were to be invited. Although rebuffed again and again, via the repeatedly stated views held by the astronomer royal and other members of the advisory committee established by the lords of the Committee of Council on Education, the Italian government's efforts continued through 1891, its ambassador again and again imploring the British government to reconsider its position.[42]

Even the IGA weighed in. In September 1890 its Permanent Commission responded to the fact that several of its member specialists had been consulted by their respective governments concerning the value of Jerusalem as the initial meridian and the usefulness of considering both this use and that of a universal time at a special conference to be held in the near future. The committee declared that "not one reason existed to change the resolutions on the subject adopted in 1883 during the International Geodetic Conference at Rome, and whose essentials had been ratified by the very large majority of States represented at the Washington diplomatic conference in 1884."[43]

Additional discussions regarding the prime meridian took place at the Fifth International Geographical Congress, which met in Berne in August 1891.[44] A special session was held so that delegates could consider the question of an initial meridian and a universal time once more. Little if anything substantial was accomplished, and the subsequent remarks of the delegate from the Indian government are a fitting close to the subject:

The meridian question, although it is apparently as far from solution as it was previously to the Washington Conference, has certainly advanced far enough that all English maps should possess a common origin for longitude. At present this is not so, for maps of India and of parts of the bordering countries are published with a longitude value . . . differing about two and a half minutes [in arc] from the true Greenwich value. . . . [S]ince attending this Congress I have come to the conclusion that a continuance of the present system is a grave disadvantage if we wish to persuade other nations to adopt Greenwich as the longitude origin.[45]

TABLE 6.2

Initial meridians on topographic maps, late nineteenth century

Country	1885	1898
Austria-Hungary	Ferro	Ferro
Belgium	Brussels	Brussels
Denmark	Copenhagen	Copenhagen
France	Paris	Paris
Germany	Ferro	Ferro
Bavaria	Munich	Munich
Great Britain & Ireland	Greenwich	Greenwich
Italy	Rome	Rome
Netherlands	Amsterdam	Amsterdam
Norway	Ferro, Christiana	Ferro, Christiana
Portugal	Lisbon	Lisbon
Russia	Ferro, Paris, Pulkova, Warsaw	Pulkova
Spain	Madrid	Madrid
Sweden	Ferro, Stockholm	Ferro, Stockholm
Switzerland	Paris	Paris
United States	Greenwich, Washington	Greenwich, Washington

SOURCE: Data from U.K. War Office, *Notes of the Government Surveys* and *Recent Large Scale Maps*, and Wheeler, *Report upon the Third IGC.*

Despite all the evidence documenting its failure (see Table 6.2), the International Meridian Conference's passage into history left in its wake two myths regarding its impact: that it signaled the formal acceptance of Greenwich as the world's initial meridian by the twenty-five "civilized" nations whose representatives had deliberated there, and that it established the adoption of world time zones based on Greenwich. Nothing could be further from the truth.

CHAPTER 7

Altering the Astronomical Day

It is much to be lamented that Seamen and Astronomers should not reckon their days alike, as this difficulty causes much confusion, especially amongst younger hands in looking for, and taking things out of the Nautical Almanac.
—William Wales, HMS *Resolution*, 1775, in J. C. Beaglehole, *The Journals of Captain Cook on His Voyages of Discovery*, 1961

Uniformity in . . . all matters relating to time . . . is of such vast and paramount importance . . . as to outweigh every consideration of technical convenience or custom.
—Sir John Herschel, *Outlines of Astronomy*, 1849

[T]he Conference expresses the hope that as soon as may be practicable the astronomical and nautical days will be arranged everywhere to begin at mean midnight.
—Delegates to the International Meridian Conference, 1884

Innocuous phrases often create firestorms of controversy; the above-quoted Resolution VI of the International Meridian Conference is one such example. The intensity of the conflict and its spread throughout the world community of astronomers prevented consideration of all other conference resolutions. Years after the firestorm had burned itself out, advocates for altering the astronomical day poked at the cold ashes, hoping to find signs of new growth. Their effort failed, leaving a long, sad history of a group of professionals led by Sandford Fleming who were unable to take no for an answer.

The printed record indicates that Resolution VI's introduction and its acceptance by meridian conference delegates were simply an early warning for astronomers "to begin to make the changes . . . which may be necessary for seamen" now that the start of the universal day was being linked to the civil day rather than to the reckoning of the astronomical one.[1] And though the nautical day was also named in the resolution, citing it was probably unnecessary. In the fall of 1805 the lords commissioners of the Admiralty had ordered that entries in the logbooks of all Royal Navy ships be given in terms of the civil day, with the U.S. Navy making the identical change by 1848.[2] Although the century-old complaint of Captain James Cook's astronomer William Wales was still valid, the onus to act was on the astronomical community.[3]

The firestorm's kindling was lit on 4 December 1884, barely a month after the meridian conference concluded. Applying the match was Commodore Samuel R. Franklin (1825–1909), superintendent of the U.S. Naval Observatory, who had been a conference delegate.[4] On the very day that President Chester A. Arthur was transmitting the official proceedings of the International Meridian Conference to the House of Representatives, Franklin issued General Order No. 3, which directed "that on or after the 1st of January, 1885, the astronomical day shall be considered as beginning at midnight, corresponding to the civil date" (see Figure 7.1).

The following day Franklin informed Simon Newcomb (1835–1909; see Figure 7.2), superintendent of the U.S. Navy's Nautical Almanac

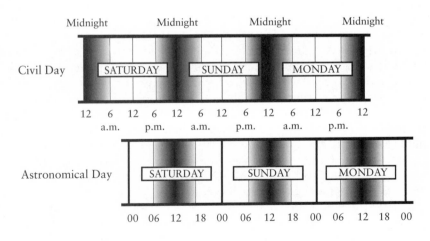

FIGURE 7.1

The civil day, which begins at midnight and is 24 hours in duration, is often split into two 12-hour sections. For centuries the astronomical day, also 24 hours long, began at noon after the commencement of the civil day.

FIGURE 7.2

Simon Newcomb, ca. 1885. The astronomer was opposed to altering the astronomical day and from 1885 on mounted a successful campaign against doing so. USNO Library photograph collection.

Office, of the impending change in recording all observations made at the Naval Observatory: shifting their reckoning from noon to midnight was being done "in accordance with the recommendation of the International Meridian Congress" [*sic*]. Immediately Newcomb, considered by many to be the most important American astronomer of the late nineteenth century, drafted a communication to Secretary of the Navy Chandler.[5] In it he detailed the impacts that a change in the reckoning of astronomical time would have on the literature of astronomy, on its teaching, and on the published tables in the *Nautical Almanac*—impacts that would extend far into the future. Newcomb recommended that no change be made in the reckoning of time in any Nautical Almanac Office publication until there was international agreement among nations on the date to make the change. A few days later Newcomb's letter to the secretary was sent to Franklin at the Naval Observatory together with his immediate superior's request for comments.

All evidence indicates that Franklin was completely unaware of the impact that a change in the reckoning of the astronomical day would have on the U.S. Navy's *American Ephemeris and Nautical Almanac*, for the Nautical Almanac Office was not part of his responsibilities. It seems fair to suggest that his knowledge of the close scientific links between the sister astronomical organizations was spotty at best; indeed, as one historian recently observed, Franklin's qualifications for the

Naval Observatory assignment were "high rank and great reputation."[6] Consequently, he was hard pressed to justify his unilateral decision.

Franklin did his best, however. He cited the Naval Observatory advisory board he had created;[7] invoked the name of the British astronomer and his fellow conference delegate John Couch Adams, who favored this change; and reminded his immediate superior that the recommendation to alter the start of the astronomical day had been unanimous—not mentioning, of course, the tiny number of astronomers who had actually participated in the meridian conference deliberations.

Franklin then stated that Astronomer Royal William Christie (see Figure 7.3), his counterpart as director of the Royal Observatory at Greenwich and an early advocate of change, had also decided to shift the beginning of the astronomical day. Moreover, he declared, the British astronomer was planning to do this on the very same date that he, Franklin, had designated for the U.S. Naval Observatory's change.[8] "[T]he fact that the first and most conservative observatory in the world had acceded to the proposal of the Conference would seem to be a sufficient reason why we should not wait for future developments," he argued. (Undoubtedly Franklin was dismayed to learn soon after that the astronomer royal planned to institute the change only internally while "awaiting official communication before introducing [the shift] generally"; that is, prior permission from the Admiralty, which was not forthcoming, would be

FIGURE 7.3

Astronomer Royal William H. M. Christie, ca. 1881. From 1883 on Christie publicly advocated altering the astronomical day to conform with the Greenwich civil day. Mary Lea Shane Archives of the Lick Observatory.

required to shift the Royal Observatory's publications of its astronomical observations to Greenwich civil time.)[9]

Arrayed against Franklin's position, however, were not just Newcomb's concerns, but also the fact that President Arthur's transmittal of the meridian conference's proceedings to Congress required a response in the form of legislation. This meant that no executive action would be taken until the legislators had completed their deliberations.[10] Franklin had no choice. He suspended until further notice the execution of the Naval Observatory's General Order No. 3, saving face by announcing his intention to query the American astronomical community "to ascertain their views on the subject."[11]

Eleven American astronomers responded to the Naval Observatory's circular. The great majority of them favored shifting the beginning of the astronomical day on 1 January 1885, in unison with the Royal Observatory at Greenwich. Four astronomers, however, pointed out that the American naval almanacs for the years 1885, 1886, and 1887 were already in print and that their tabulations would be in conflict with a shift in the reckoning of the astronomical day. A very few also wrote that it would be best to postpone any change until there was general agreement among the world's astronomers, not just among American ones. None of these outside views carried any weight, of course; only Newcomb's position mattered.

Thus on 29 December Franklin once again suspended the execution of General Order No. 3, announcing that the Naval Observatory "will make no change in the beginning of the astronomical date, the weight of opinion being that it will be better to defer such a change until the ephemerides are constructed in accordance with the recommendation of the recent International Meridian Conference." The following February, Secretary of the Navy Chandler submitted the various communications, which were referred to the Senate's Committee on Foreign Relations.[12]

Although Franklin had extricated himself and the Navy Department from what could have been a rather awkward situation vis-à-vis Congress, the fire continued to spread. Newcomb added fuel via a letter printed in the 9 January 1885 issue of the Royal Astronomical Society's *Monthly Notices*,[13] in which he applauded both the creation of a universal day and its link to Greenwich civil time, predicting that it would be a "great convenience . . . [i]n the reckoning of physical and meteorological phenomena." Then, turning to his real concern—that the astronomical day would be shifted to conform—he further asserted that lack of concordance between the reckoning of universal time and astronomical time was not a serious problem. "I can recall no instance of trouble or confusion in the past resulting from the difference between the astronomical and the civil day," Newcomb wrote—a rather surprising statement coming from someone in charge of preparing predictive tables for sailors and one that suggested the astronomer was unfamiliar with their practical problems.[14]

Quite correctly, for the recommendation only expressed a hope for change, Newcomb insisted that no strong reason had been presented at the International Meridian Conference for shifting the reckoning of the astronomical day. Voicing his concern that those astronomers in favor would simply shift the reckoning of their own observations, which would lead to great confusion, Newcomb urged that nothing be done until a common agreement among astronomers existed. Further, he concluded, should it eventually become necessary to change the reckoning, "The beginning of the coming century would be a very good epoch, and would allow about the right time for consideration."[15]

Two weeks after Newcomb's views appeared in England, Wilhelm Foerster (1832–1921), director of the Berlin Observatory, added more fuel via an article in *Astronomische Nachrichten*.[16] Foerster, an invited delegate at the IGA conference in Rome, insisted that the universal day and the astronomical day of Greenwich should coincide, with the former also starting at noon. He saw no reason why a change in date should not occur at noon in Europe, for his view was that the universal day was of no interest or use to the general public, the railways, or the telegraphs and other communications links. He added that prominent and powerful German astronomers were also opposed to the change.[17]

Foerster's remarks were followed by those of the Viennese astronomer Theodor von Oppolzer, who wrote to rebut Newcomb's views regarding the confusion that a change in astronomical reckoning would engender.[18] At Rome, Oppolzer, one of the secretaries of the IGA's Permanent Commission, had voted in favor of having the universal day and the astronomical day conform, the majority view among delegates. Now he strongly favored the recommendation adopted at Washington: to have the universal day conform to the civil day at the Greenwich meridian. Oppolzer quoted Sandford Fleming's view that before long and throughout the world the general public would adopt universal time—Greenwich civil time—for their activities, but this radical view was never accepted.

Of greater significance, Oppolzer also argued that those astronomers who retained astronomical time in their work would be in conflict with those who relied most on their publications: seamen who now used the civil day in their logs and those individuals preparing "calendars and other publications destined for general use." He reminded his colleagues that for solar and lunar eclipses tabulated in astronomical reckoning, historians often found that a particular event was given in the written record one day earlier than the day predicted by the astronomer—one more demonstration that the ephemerides astronomers prepared were not simply for their use alone, as Newcomb argued. Oppolzer's views later led him to prepare his famous *Canon of Eclipses* in terms of the universal time definition recommended by the International Meridian Conference.[19]

Oppolzer was not the only astronomer who favored shifting the origin of astronomical time. I have already mentioned Astronomer Royal William Christie and the director of the Cambridge Observatory, John Couch Adams. Another was Otto Struve, who in the spring of 1885 published a lengthy summary and analysis of the International Meridian Conference's recommendations.[20] This longtime advocate of the Greenwich meridian traced the need for uniform time in the sciences, commenting that astronomers and, surprisingly, navigators had the least need of all for uniformity, for "the navigator, like the astronomer, has continually to consider the principle of difference of Time, and hence a law relative to the Unification of Time notation is of less relative significance to him."

Struve contrasted the differing results at the Rome and Washington conferences with regard to the commencement of the universal day. He suggested that the difference was due in part to the absence of seamen at the former when astronomers were arguing the navigator's need, and the authoritative statements made by men of the sea at the latter. In his view the astronomers, who desired to have the universal day begin at midday, were the only group out of step. With perhaps a touch of sarcasm, the distinguished astronomer and geodesist wrote that when the universal day and civil day at Greenwich conformed, astronomers could "quietly remain in their old customs without grieving themselves as to the arrangement of a matter indifferent to them." He scolded his fellow astronomers by noting that the mission of state-supported observatories was to serve not just scientific purposes, but also practical ones. That meant they should engage in active communication with various groups in the public sector, whose lives were ordered in terms of the civil day.

Struve predicted that the astronomers' use of a different system of dating would eventually have to be abandoned. Based on the long lead times involved in preparing nautical almanacs and other predictive tables, he recognized that no change could be implemented before 1890. As director of the Central Astronomical Observatory at Pulkova, his own preference for change was the turn of the century. And thus one of the giants of nineteenth-century astronomy and geodesy supported the recommendation of the diplomatically oriented International Meridian Conference.[21]

Struve's remarks did not end debate, however. Other astronomers began to be heard. One or two criticized Christie for his decision to publish in advance of a common agreement even a part of the Royal Observatory at Greenwich's output in terms of civil time. Others voiced their opposition to any change at all.

By mid-April the American journal *Science* was lamenting the likely failure of the International Meridian Conference recommendations and blaming the controversy surrounding the hoped-for alteration in the astronomical day as the main culprit. A week later, reflecting the navy's continuing efforts to have those recommendations implemented and hoping to break the impasse, the recently confirmed Secretary of the Navy, William Whitney, asked the National Academy of Sciences to consider the matter of shifting the beginning of the astronomical day in the *American Ephemeris and Nautical Almanac*.[22]

Late in June an American astronomer reporting on the controversy concluded that it was difficult to see how the matter would be decided, having counted thirteen colleagues in America and abroad in favor of the change and fourteen opposed. He did not weight their status, however, which would have led him to a more definitive conclusion. For one of the opponents (Newcomb) was directly responsible for his government's almanac, the second (Foerster) had already indicated extremely strong opposition in his country, and the third opponent of change was the president of the Astronomische Gesellschaft.[23]

The Astronomische Gesellschaft was an international organization founded in 1863 to advance science via long-term cooperative programs. Over three hundred astronomers and other scientists were members; its highly regarded stellar catalogs were in use throughout the world. That August the society held its three-day biennial meeting in Geneva, with forty-one members participating in three formal sessions. One session focused on the astronomical day.

The president of the society, G. F. Arthur Auwers, opened the session by announcing that the society's executive committee had already discussed the issue and stood seven to one against any change, that the subject had been placed on the agenda to allow members to voice their individual opinions, and that a vote would not be taken.

Past president Otto Struve spoke first, his position no different from what he had expressed the preceding spring in his printed remarks. Simon Newcomb repeated his reasons for opposing the change. Other key astronomers addressed the session; all favored maintaining the status quo. One was Adalbert Krueger, who stated that he would not introduce the new time calculation into the *Astronomische Nachrichten* until a great majority of astronomers were in favor of doing so. Another was Friedrich

Tietjen, editor of the *Berliner Astronomisches Jahrbuch*, who announced that he would make no change in the beginning of the astronomical day in his volumes. At the end of the session the general view was summarized by Auwers, who considered the issue an internal matter, not to be decided outside the astronomy community. Moreover, should the change ever be agreed to, it must be made by all observatories at the same time.[24]

An early reporter of the August meeting of the Astronomische Gesellschaft wrote that "the proposal for changing the reckoning of Astronomical time was almost unanimously rejected." The Swiss astronomer Étienne Gautier, who summarized the session soon after, expressed bewilderment at the intensity of the debate, noting that another International Meridian Conference resolution had given astronomers an out: that the universal day was for purposes "for which it may be found convenient, and which shall not interfere with the use of local or other standard time where desirable." How could a simple hope expressed in another resolution end up causing acrimony sufficient to injure one of the Astronomische Gesellschaft's principal goals, cooperation among scientists? "We cannot explain it," he lamented.[25]

Of course this August session of the Astronomische Gesellschaft did not end the controversy. At the end of the year the National Academy of Sciences submitted its report to Whitney. The members of the committee who drafted it were much in favor of shifting the beginning of the astronomical day and consequently minimized the intensity of opposition at the Geneva meeting by writing that the issue "was considered at some length and discussed at some length, but was left undecided." This unsupportable statement was included presumably because no formal vote had been taken.

Fortunately, these academy members' public support for change did not go so far as to recommend unilateral action on the part of the United States. Rather, they recommended that "the change should be made as soon as sufficient concert of action can be secured among the leading astronomers and astronomical establishments of the civilized world." Because that very same decision had been articulated by the superintendent of the Naval Observatory over a year earlier, one suspects that Whitney may have regretted ever asking the academy to consider the matter.

In any event, the academy's brief summary of its recommendation ended up completely obscured, for the report's main thrust—a new facility for the Naval Observatory—overwhelmed it. Moreover, when the now-thirty-six-page document was sent to the Senate, it was referred to the Committee on Naval Affairs—quite the "wrong" legislative committee to consider any alteration of the astronomical day.[26]

Although the academy's specific recommendation had been effectively buried, that did not prevent the Naval Observatory astronomer Asaph Hall from launching an assault. Writing in *Science* a month later, he held that contrary to general opinion, academy reports did not necessarily reflect the views of the majority of its members. In particular, this one represented "only the opinion of the committee who sign[ed] the report," and probably "a majority of the astronomers of the academy would oppose such a change *if they were permitted to speak*" (emphasis added). This remark implied a dictatorial environment within this august body of specialists.

The following month the *Observatory* printed Hall's list of seven astronomers associated with the Naval Observatory who were opposed to a change in the beginning of the astronomical day. Continuing his attack, Hall asserted that their opinions were "of greater value than that of the admirals, generals, and diplomats who so largely formed the [International Meridian] Conference"—once again espousing a naive view of the role of specialists in setting overall government policy.[27] Nonetheless, Hall's remarks certainly typified the feelings of many astronomers regarding non-astronomers' daring to express even the hope of an alteration in "their" day.

At this juncture, the matter of altering the basis of astronomical time should have ended. However, seven days after Whitney submitted the National Academy of Sciences' report to Congress, the secretary of state, T. F. Bayard, received a note from the British ambassador. It contained a query approved by Her Majesty's secretary of state for foreign affairs: is the United States, as convener of the International Meridian Conference, ready "to invite the adhesion of other maritime States to the sixth resolution of that Conference"?[28]

The British inquiry had been a while in coming. In 1885 the advisory committee established by the lords of the Committee of Council on Education recommended that various government departments and learned societies be consulted regarding actions that should be taken as a result of the International Meridian Conference. Among those who responded, the lords commissioners of the Admiralty expressed their willingness to adopt civil reckoning in the *Nautical Almanac*, "if other maritime nations are prepared to adopt the proposed reckoning of astronomical time" in their own almanacs so as to reduce the possibility of mistakes by navigators during the transition period. The advisory committee then recommended that the Foreign Office raise the issue with the United States. The several necessary actions were taken, and the inquiry regarding the sixth resolution arrived at the State Department in mid-February of 1886.

Routed through the secretary of the navy, the inquiry arrived at the

Naval Observatory along with a request for its recommendations. At once its superintendent, Commodore George E. Belknap (1832–1903), made absolutely sure that Simon Newcomb, superintendent of the Nautical Almanac Office, would participate formally in the process. This was a far cry from the actions of Belknap's immediate predecessor, Franklin, whose unilateral decision had done so much to ignite the controversy.

Belknap and Newcomb together began preparing the formal response. Belknap also requested the views of the USNO's five professors of mathematics. All were opposed to altering the start of the astronomical day, the common reason cited being confusion among astronomers.

However, one respondent focused on how the issue affected practical astronomy, the basis of ocean navigation—the rationale for the Naval Observatory's programs and an area essentially ignored throughout the astronomers' debates. Noting that no practical navigator had asked for a change in the reckoning of the astronomical day, he argued that the current system was simple in practice and had led to no confusion. Further, a change would require alteration of the fundamental definitions of nautical astronomy, and all textbooks would have to be rewritten. Of course, the professor of mathematics understood that after such an alteration had been adopted, opposition to it would slowly die out. Nevertheless, he was opposed to such a change, "at least until the matter could be brought to the attention of practical navigators."[29] His sensible suggestion was ignored (hardly surprising given the short period of time allotted for responding to high-level department inquiries).

The Navy Department's formal response, prepared by the superintendents of the USNO and the Nautical Almanac Office, was an excellent bureaucratic reply. It began with the valid point that the International Meridian Conference's sixth resolution was tied not to the preceding one—a universal day based on the civil day at the Greenwich meridian—but to "*local* astronomical . . . days, as used by astronomers in their several observatories" (emphasis added).[30] It noted that if the resolution were to be adopted, "Washington astronomers will count their days from Washington midnight, the Berlin astronomers from Berlin midnight, &c." Adopting the sixth resolution would not foster unification among astronomical tables and almanacs, nor would it lead to a common standard of time, one of the two primary objects of the International Meridian Conference.

The two superintendents then contrasted the conference's first five resolutions with the sixth and seventh ones. They termed the first group "proposals," all of them in keeping with the object of the president's invitation to the nations of the world. The latter two they called "simple expressions of hope," to be judged simply on their own merits. Here they noted the opposition of key German astronomers, summarized the debates at the August meeting of the Astronomische Gesellschaft, and speculated that the French would resist such a change.[31]

The respondents also cited the survey of American astronomers undertaken at the end of 1884 by the USNO superintendent at the time, Franklin, and the recent National Academy of Sciences report. While conceding that astronomers in England were generally in favor of change, they countered that even if most astronomers supported or would accept the altered reckoning of the astronomical day on the condition that others did, "the majority of the official astronomers, who are most immediately and practically interested in the problem, are against its introduction," noting the unanimity among the U.S. Navy's professors of mathematics, whose remarks they attached to their report. (They also attached Newcomb's December 1884 letter to the secretary of the navy outlining the great inconvenience that such a change would cause.) Based on the superintendents' analysis, three of the "big four" producers of astronomical ephemerides—France, Germany, and the United States—opposed altering the beginning of the astronomical day.

For a general reader, such a verdict would likely be considered sufficient to end further discussion of the subject. But Belknap and Newcomb were longtime and successful manager-bureaucrats in their respective spheres, so they continued their detailed analysis. They noted that the world's important observatories had been linked via the electric telegraph, and thus their longitude differences were known with extreme precision. Consequently, it would be quite easy to make a unifying change to universal time in the products of these observatories.[32] Since this approach was not mentioned at the International Meridian Conference, the superintendents deftly implied the delegates' lack of expert knowledge by writing that "we can scarcely doubt that the failure of the Conference to recommend it arose from the circumstance that the [telegraphic longitudes] subject was not called to its notice."

Reminding readers that they had been directed to consider the sixth resolution of the International Meridian Conference and no others, Belknap and Newcomb speculated on why the British government had not included consideration of the two universal-day resolutions in its official query to the United States and asserted that "[s]o far as astronomers are concerned, it is . . . a matter of indifference whether their data are referred to noon or to midnight of the Greenwich meridian." Those in the navy and the merchant service had called for no such change. The current system worked well and was embedded in all textbooks.

In conclusion, Belknap and Newcomb stated that in their judgment, change would "lead to no little confusion, and sometimes disaster, before navigators became accustomed to the new method and its requirements." They judged the typical navigator in the merchant service to be a person accustomed to routine work and dependent on rules of thumb—certainly not a skilled mathematician or astronomer able to adjust to change with ease. So the debate had finally come full circle: there should be no change in the astronomical day because navigators would be disadvantaged!

The report's summary paragraphs displayed the authors' great skill in bureaucratic writing. First the two officials summarized their position:

In thus pointing out the inadvisability of making any change in astronomical ephemerides, we do not conceive ourselves recommending any action adverse to the recommendations of the Meridian Conference, or to the proposal of those recommendations by the president [of the United States] to foreign nations. So far as the universal day is concerned . . . the Conference did not recommend its adoption in the Nautical Almanac[s], but very wisely restricted it to cases in which it should be found convenient. We do not deem it convenient to introduce it into astronomy at present.

Then they ended the report with these words:

But if, despite the views and considerations herein expressed, this Government, as the convener of the Prime Meridian Conference, deems it best to adopt and conform to the suggestions embodied in the sixth resolution of the Conference, then the undersigned beg leave to suggest that the date of 1900 may be fixed upon for such purpose.

In mid-April 1886 the secretary of the navy forwarded Belknap and Newcomb's report to the secretary of state, who then transmitted the "voluminous correspondence" he had received to the British government. None of these documents contained a direct answer to Great Britain's inquiry respecting American readiness to invite the adhesion of other maritime countries to the sixth resolution of the International Meridian Conference; apparently it was deemed unnecessary to do so. Altering the start of the astronomical day, already a dead issue in the United States, had received an official obituary.

In retrospect, it seems a pity that the delegates to the International Meridian Conference ever expressed themselves on this subject via their essentially pointless declaration. For the truly grand and useful ideas embodied in other resolutions—a prime meridian, a universal time—fell by the wayside as astronomers hardened their positions with respect to keeping "their" own time. Given the strong rebuttal by the American government's astronomers, nothing could or would be done by the executive branch, much less by Congress. Indeed, the nineteen months of debate had resolved only one question: namely that all four governments producing the world's major nautical almanacs—France, Great Britain,

Germany, and the United States—must agree before any alteration could be discussed, let alone effected.

In 1892 Sandford Fleming took up the cause once more. Why he did so at this particular time is not known. Perhaps it was the result of his very modest success in advancing other aspects of uniform time despite a decade and a half of proselytizing. He had played scarcely any role in the North American railways' adoption of Standard Railway Time. His efforts to promote the twenty-four-hour notation had led to its use by a grand total of two Canadian railways and the ASCE, with no sign that others in North America were contemplating it. His attempts to influence legislators by drafting a bill for Canada's Parliament to legalize the railways' uniform time—as well as working to have similar bills introduced in both houses of the U.S. Congress—had failed.[33] And though the world's adoption of time zones, in which Fleming had begun to play a role, was advancing, others deserved far more credit for this than he did.[34]

Fleming began his campaign early enough so that an altered astronomical day could be in place by the start of the new millennium. Moreover, he did not stint in what would be his last attempt to extend time uniformity, enlisting professional societies, the governments of Canada and Great Britain, and a host of individuals. He forged an alliance with Charles Carpmael (1846–1894), superintendent of the Meteorological Service of Canada and a recent president of the Canadian Institute, which led to the formation of a joint committee of the Canadian Institute and the Astronomical and Physical Society of Toronto (now the Royal Astronomical Society of Canada), with Fleming as chairman. The committee's charter was "Astronomical Time Reckoning."[35]

In April 1893 the joint committee addressed an eleven-page circular-letter to the world community of astronomers asking for their position on changing the start of the astronomical day at the turn of the new century; a ballot was included. About one thousand circulars were distributed. The response was disappointing, as even Fleming admitted: only 171 replies. Nevertheless, 108 individual astronomers were in favor of the change while 63 were against, a ratio sufficient for the proselytizer to assert that the great majority of the world's astronomers supported the change. Fleming also analyzed the individual responses in terms of country-of-origin, finding that astronomers residing in twenty-three countries were in favor, those in four countries were against, and astronomers in one country were split fifty-fifty.[36]

Fleming then tackled what was for him the fundamental issue: to have the British *Nautical Almanac* altered so the astronomical day would conform to the civil day. Mimicking Fleming's 1879 campaign, the joint committee now sent its report directly to the governor-general of Canada. Along with it went a memorial asking that the materials be transmitted to the home government for its consideration and, hopefully, action. The usual process took place, and in June 1894 the colonial office sent copies of the Canadian materials to the Admiralty and to the Science and Art Department for their comments.[37]

The Admiralty's response was a negative one, the lords commissioners of the Admiralty regarding the "list of names of Astronomers and other gentlemen who have given an opinion" as not authoritative, for it omitted "nearly the whole of the names of leading Astronomers in the world." Moreover, they repeated what they had written in 1886: "that until their Lordships receive the official views" of the other eight countries that published astronomical ephemerides, "they are unable to consider the question of authorising alterations of the Nautical Almanac in the manner suggested."[38] (Austria, Brazil, Mexico, Portugal, and Spain had now been added to the "big four" of almanac producers.)

The response from the Science and Art Department was no more encouraging. Its standing committee, many of whose members had advocated both the Greenwich meridian and a universal day conforming to the civil day, echoed the Admiralty, noting how few names of leading astronomers appeared in the Second Report of the Joint Committee. Wary of directly advocating an alteration of the Admiralty's *Nautical Almanac* but still anxious to assist in effecting change, it recommended that the British government raise the issue with foreign governments. Accordingly, in September the Foreign Office issued instructions to its representatives in the eight affected countries. In the transmittal letter sent to the United States requesting its views, it was noted that although the lords of the Admiralty "do not consider the change necessary, they are nevertheless prepared to carry it out in 1901, provided that other nations who publish astronomical ephemerides desire the change and will take the same action."[39]

On receiving the British government's letter, the U.S. State Department forwarded the inquiry to the Navy Department for its views. Three senior members of the USNO—the superintendent, Captain F. V. McNair, and the mathematics professors Simon Newcomb and William Harkness—prepared the department's response. Their words became the U.S. government's official position: "[W]e are decidedly opposed to any change in the existing mode of reckoning astronomical time, and therefore recommend that no departure be made from the present system." Forwarded to the British government's representative in Washington, the secretary of state, W. A. Gresham, described it as "adverse to the Canadian proposition on the subject in question."[40]

Hearing rumors that the United States had responded negatively, Fleming contacted his longtime uniform-time associate, Cleveland Abbe, and wrote: "If this [negative position] be taken as the final decision of the United States it certainly means that the reform will never be made in any part of the world," adding, "I cannot bring myself to believe that a country so advanced as the U.S. is in everything, will in this matter stand in the way of a reform proposed at the Conference by U. States delegates and now after years of delay assented to by the most conservative nations of Europe." He asked Abbe for suggestions on how to obtain "a reconsideration of the subject by the U. States" and also for his assistance, suggesting a memorial from the AMS to the president of the United States as a possible action.[41]

Aware of the negative response sent by the State Department, Abbe advised Fleming that it would not be productive to try and force the editors of the *American Ephemeris and Nautical Almanac* to adopt the change. Indeed, any appeal to higher authority would simply be forwarded to the Nautical Almanac Office, where the official reply would be drafted. In addition, Abbe questioned how seriously the navigators of the world truly desired exclusive use of the civil day, remarking that astronomers and nautical almanac producers were not much troubled by the existence of a dual system of time. Moreover, "[I]f they wish to keep it up in their observatories and ephemerides and astronomical tables we, outsiders, can hardly prevent it." In Abbe's view, the only feasible way to effect change was via a lobbying effort, one involving "Chambers of Commerce, Boards of Trade, steamship and transportation companies[,] marine insurance companies and underwriters[,] and all others interested in ocean navigation." In other words, if the astronomical day was to be altered, pressure must come from the concerned private sector.[42]

Heeding Abbe's advice, Fleming embarked on the next phase of his campaign, enlisting W. Nelson Greenwood of Lancaster, England, publisher of *Greenwood's Nautical Almanac* and other aids for navigators.[43] In mid-1895 Greenwood distributed a circular letter to shipmasters entering major British ports in which he solicited their opinions with respect to replacing the astronomical day with the civil day; his results were published several months later. To the question "Are you in favour of the Unification of Time as applied to the Civil, Nautical and Astronomical Days, and is it desirable in the interests of all concerned that such days should commence at mean midnight?" 96 percent of the 243 responding shipmasters were in favor.[44]

Then a severe blow to Fleming's campaign came in the form of correspondence from the home government relaying the results of its official inquiries, which documented a less-than-unanimous willingness

to alter the astronomical day: six countries were in favor, one was opposed (United States), and two more (Germany and Portugal) did not
respond.[45] Transmitted via the governor-general of Canada, the materials
were referred to the societies' joint committee for its opinion.

At around this time (the summer of 1895) the campaign to alter the
beginning of the astronomical day began to take on a surreal quality.
In the joint committee's Third Report, "Unification of Astronomical,
Civil and Nautical Time," Fleming, as chairman, wrote that a "brief
communication had been received from the Secretary of State in Washington," one that included a copy of the U.S. Naval Observatory's report.
After expressing great surprise, he stated that its view was "entirely at
variance with the position taken by the United States throughout the
movement for reforming the time-reckoning of the world during the last
fifteen years," concluding, "With all the facts before us, it is impossible
to consider that the adverse report signed by three officials of the United
States Naval Observatory, fairly represents the mind of the United States
Government, of Congress, or of the people of the United States." U.S.
Secretary of State Gresham must certainly have been astonished to read
this conclusion!

Fleming then began a lengthy personal attack on Newcomb, citing
selected comments made more than a decade earlier and not germane to
the current issue. Fleming claimed that 99 percent of the American and
Canadian civil engineers who replied to his 1882 survey for a uniform
time for North America expressed opinions diametrically opposed to
those of the astronomer. (Of course, he did not point out that Newcomb
himself had actually voted in favor of a uniform time for the United
States.) Then he turned to the joint committee's 1893 survey of astronomers, emphasizing those residing in the United States, a group in favor
of the shift by twenty-eight to ten. (Of course, Fleming did not mention
the Admiralty's low opinion of the survey's authoritativeness.) He concluded, "In view of the facts narrated, the Joint Committee respectfully
conceives that it is fully warranted in the opinion that the United States,
as a nation, may be truly considered to be one of the nine ephemerides
publishing nations in favor of the proposal to bring the Astronomical
Day into agreement with the Civil Day."[46] Via his "analysis," Fleming
arbitrarily negated the U.S. government's official rejection of his proposition, thus enabling him to count it with the non-respondents Germany
and Portugal as expressing no formal dissent.[47]

This third report of the joint committee was also sent to the home
authorities by the governor-general. On 12 December 1895 the secretary
of the Admiralty responded with, "I am directed by their Lordships to
request that you will inform the Secretary of State [for the Colonies] that

as the condition under which they consented to consider this question [unanimity among almanac producers] . . . has not been attained, my Lords have no intention of moving in the matter."[48]

Incredibly, the campaign still went on. The following spring the secretaries of the Canadian Institute and the Astronomical and Physical Society of Toronto sent a joint communication to the governor-general enclosing the first summary of Greenwood's survey of shipmasters, information they considered so important that they asked that the governor-general forward it to the home authorities "in the hope that the subject of Uniform Time may be *reconsidered*" (emphasis added).[49] He did so, but to no avail; the Admiralty eventually replied that this latest Canadian communication contained nothing that would "affect their Lordships' decision as contained in Admiralty letter of 12th December 1895."[50]

The Royal Society of Canada now entered the campaign. Seemingly starting afresh, it presented a memorial to the governor-general asking him to take steps "calculated to secure the adoption by Her Majesty's Government of the sixth resolution of the International Meridian Conference of 1884." Next, the Society's council elected to communicate on the subject with the boards of trade of important British and Irish cities and towns—here, one suspects Cleveland Abbe's lobbying suggestion had been taken to heart.[51]

In October the society's advocacy document, "The Unification of Time at Sea," was finished. Over forty pages long with seven appendices, it briefly summarized the history surrounding the International Meridian Conference, asserted—wrongly—that most of the resolutions adopted were now generally accepted throughout the world, and focused on the sixth resolution: changing the reckoning of the astronomical (and nautical) day.

After stating that the Admiralty favored such a change but would not act until all nations that prepared ephemerides were in agreement, the author argued that unanimity in such matters was probably not attainable—an admission that should have immediately ended the discussion. Instead, the author paraphrased Sandford Fleming's conclusions—"that seven of the nine ephemerides publishing nations may be considered assenting parties to the proposed change"—adding that the two non-respondents "have in no way expressed dissent, and that their silence . . . may fairly be taken as *equivalent to concurrence*" (emphasis added). That this document was signed by no less a personage than the honorary secretary of the Royal Society of Canada, clerk of Canada's House of Commons, and adviser to its leadership on parliamentary and constitutional matters is simply astonishing.[52]

At its next annual meeting (June 1897), the Royal Society of Canada reported communications from the governor-general of Canada to the secretary of state for the colonies, formal efforts by a host of chambers of commerce seeking to sway the British government, similar activities by the association of Lloyd's, a memorial from the four-thousand-member Royal Colonial Institute to the British prime minister, and the appointment of an influential committee by the Royal Society, London, to consider how best to intercede on behalf of its Canadian counterpart.[53]

The crushing blow came in a letter from the Royal Society, London. British astronomers were far from solidly in favor of change (sixteen to four), as asserted in the joint committee's survey; the Royal Society's own committee found "a great diversity of opinion as to the advisability of civil reckoning for astronomical purposes." Moreover, because the change could not be effected in the 1901 *Nautical Almanac*, committee members recommended that the London society take no steps to support the proposed change. The Royal Society's secretary informed his Canadian counterpart that the committee's report and recommendation had been accepted by the council, and that despite the Royal Society of Canada's request, the society would not attempt to influence Her Majesty's Government. Further, the letter concluded, it was not now opportune for it to bring the issue to the "attention of the powers who have not already assented to the change."[54]

Still grasping at straws, in late June the Royal Society of Canada passed a resolution requesting the BAAS to cooperate with it and with other Canadian societies in attempting to influence Her Majesty's Government. In August the president of the Astronomical and Physical Society of Toronto addressed the Physical and Mathematical Section of the BAAS during its annual meeting in Toronto; his subject: "On the Unification of Time." He noted the Royal Society of Canada's resolution and expressed hope that the British association would assist in bringing the subject "before the nations of the world for final consideration."[55]

As reported in *Nature*, remarks opposing the change were made afterward by Simon Newcomb, while the physicist Arthur Rücker, secretary of the Royal Society, summarized the inquiries made by the society, the Foreign Office, and the Admiralty. Extremely dissatisfied with the BAAS's position of apparent inaction resulting from these negative remarks, the officers of both the Astronomical and Physical Society of Toronto and the Canadian Institute, along with Sandford Fleming, met with members of the British association's council. The net result was the inclusion of another talk, "The Unification of Time at Sea," before the BAAS section on geography, and the passage of a rather noncommittal resolution by both BAAS sections calling the attention of the council to the subject, "in which the interests of mariners are deeply involved," and asking them "to take such action in the matter as may seem to them to be desirable." No action was taken.

Still dissatisfied and angry, the officers of the Astronomical and Physical Society of Toronto laced their report of the Toronto meeting with complaints, asserting the complete lack of independence of the BAAS from the community of astronomers, "who are members of another body, and are prejudiced against the movement, and who, the servants and not the masters of mariners, should not be consulted at all." Again it had come full circle, with the Canadian astronomers now viewing themselves as the British Empire's main supporters of the world's mariners. Still unwilling to accept the U.S. Naval Observatory's adverse report, the society's secretary argued that it represented the views of only three American officials, and, further, that "[t]he omission of the Government to approve of, or in any other way to take responsibility for the report, has been thought not to be without significance."[56] In other words, the Canadians seemed to feel that an agency document that had passed through a formal review process and then been sent through proper channels to a sovereign power might possibly still not be the official position of the United States!

But the astronomical time debate was over. In fact, the issue had effectively been decided more than a dozen years earlier. Advocate Sandford Fleming had simply refused to accept the opposition of those whose views truly mattered: the government astronomers responsible for their country's astronomical and nautical ephemerides. Marshaling all his resources and drawing on his scientific and government connections in Canada, as well as on his associates in Great Britain, he had tried to advance what he insisted was the majority view. He had failed, and finally even he had to face that fact. The unification of time was no longer a pressing matter, and the record shows no further publications by Fleming on this subject.[57]

For almost two decades no action was taken on Resolution VI of the International Meridian Conference. Not until a great passenger ship was lost on its maiden voyage and the "civilized" nations found themselves in the throes of a great war would altering the beginning of the astronomical day again become an issue of importance.

CHAPTER 8

Partitioning the World's Time

[V]ery soon no other time will be displayed in the United States than that of the Greenwich minute and second, linked to the most convenient meridian hour.
—Robert Schram, "Einheitliche Zeit," 1886

[T]he almost unanimous adhesion of the civilized countries to the meridian of Greenwich should cause the supporters of other meridians to struggle no longer on this terrain.
—Ernest Pasquier, "L'unification de l'heure," 1891

While many pursued the uniformity-postponing issue of altering the astronomical day, a separate movement was leading toward greater uniformity in public timekeeping.[1] By the start of the twentieth century, civil times based on exact hours from the Greenwich meridian were the norm throughout much of the European continent; to a lesser extent, the British Empire's colonies were also adopting times referenced to Greenwich. These national times were segments of today's worldwide system of time zones, a system that eventually supplanted essentially all other notions of public time.

In 1886 the Austrian astronomer and geodesist Robert Schram (1850–1923; see Figure 8.1) published an article titled "Uniform Time" in the *Wiener Zeitung*.[2] An associate of Theodor von Oppolzer and his eventual successor as director of the Austrian Geodetic Bureau, Schram summarized the status of universal time as discussed at the conferences in Rome and Washington. However, he emphasized the 1883 implementation of Standard Railway Time in the United States, pointing out that many of the country's cities and towns had abandoned their local times in favor of it. Schram proposed that Austrian railways adopt this system, using the time along the fifteenth meridian east of Greenwich—exactly one hour later than Greenwich time—as their operating time. Schram turned his article into a pamphlet, but his proposal, the product of an outsider insofar as railways were concerned, was ignored.

In a summary prepared some years later, Schram highlighted subsequent events. In 1888 the president of the Hungarian State Railways, who had seen Schram's writings, recommended to a conference of Austrian and Hungarian railway directors that a time one hour ahead of Greenwich be adopted as the network's common operating time. The directors accepted this proposal unanimously, and their resolution was referred to the Austro-Hungarian government's Ministry of Commerce for ratification. Approval came several months later but was coupled to a desire that the new standard be adopted by all railways within the time section: those of the German Empire, Switzerland, Italy, and Serbia. The minister of commerce indicated that the imperial government of Austria-Hungary would take preparatory diplomatic steps to foster its adoption.

In November 1889 the president of the Hungarian State Railways submitted the standard-time proposal to the German Railway Union, which appointed a special committee of delegates from the important railways of Austria-Hungary and Germany. The following January the

FIGURE 8.1

Austrian astronomer and geodesist Robert Schram, ca. 1886. His 1886 proposal that the railways adopt a Greenwich-linked time began a process that led to Central European Time throughout most of Europe. Courtesy Archives of the German Academy of Naturalists, "Leopoldina," Halle.

fifteen members of the committee endorsed the concept and unanimously agreed to propose three resolutions at the Railway Union's next general conference:

The introduction of the proposed standard time for railroad service [internal operations] is in the highest degree advisable;

The same is to be said about the use of this time for the time tables destined for the public;

The general introduction of the said standard time for everyday life is recommended.[3]

Meeting in Dresden in late July, members of the German Railway Union voted in favor of introducing the new time standard throughout the network in the spring of 1891, the usual season for schedule changes.[4]

Since many of the German railways were owned by the various states that made up the German Empire, government approval of the set of proposals was required. As for extending the new time to civilian uses, opposition appeared early on, with numerous newspaper editors against the idea, as was Wilhelm Foerster, director of the Berlin Observatory. Thus, when the proposals came before the Reichstag late in 1889, the entire set was rejected.[5]

The following year a special commission's report came before Imperial Germany's Bundesrat (Federal Council). The report recommended that the new time be introduced into the railways' exterior services as well as internal operations. Opposed by the Prussian delegates and government, the recommendation failed. The government of the State of Prussia also opposed the change.[6]

Opposition to uniformity in Germany's civil time might have continued but for an extraordinary event that took place on 16 March 1891, when budget estimates for the Imperial Railway Department, along with its views on uniform time (*Einheitszeit*), went before the Reichstag. During the session Count Helmuth von Moltke (1800–1891; see Figure 8.2) spoke. The ninety-year-old Prussian field marshal's arguments in favor of uniform time for the empire altered the situation completely.[7]

Considered by many to be the greatest military strategist of the second half of the nineteenth century, von Moltke had recognized early in his career the enormous value of a railway system for moving men and matériel and subsequently incorporated this view into Germany's war planning. At the session von Moltke, a member of the Reichstag and the recently retired chief of staff of the German army, began by noting that it was universally recognized that a uniform time was indispensable for safe and efficient railway operations.[8]

FIGURE 8.2

Helmuth von Moltke, date unknown. Count von Moltke's 1891 speech before the Reichstag urging a single time for the German Empire led directly to the adoption of uniform time linked to the Greenwich meridian. Library of Congress, LC-USZ62-31618.

In Germany however, *five* times were being employed by the railways: the local time of Berlin in North Germany and Saxony, that of Munich in Bavaria, Stuttgart time in Württemberg, Karlsruhe time in Baden; and the local time of Ludwigshafen in the Rhine Palatinate—the equivalent of five zones for a country sixty-seven minutes in breadth. Moreover, while the smaller states were using their respective railway times throughout their regions, in Prussia the railways' Berlin time was being converted to the various stations' local times for public displays and timetables (exterior services). As von Moltke reminded his fellow legislators, "This collection of times is debris left over from the era of a splintered Germany and which should be removed now that we are an empire" (see Figure 8.3).[9]

Though of little significance for the railway traveler who found the railway's time different from his watch's display at each station, "these different times become a substantial aggravation while carrying out the business of railways, especially for those services which, from a military point of view, must be demanded," von Moltke argued. Turning to the most significant issue, he declared:

[I]n the event of a mobilization, all timetables that go to the troops must be computed in both local times and in the South German standard times. Naturally,

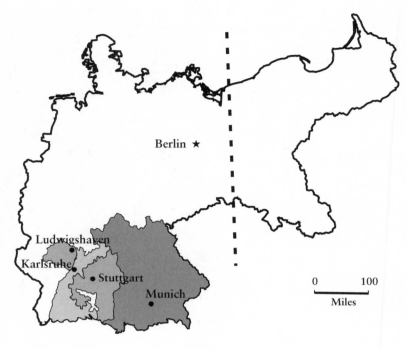

FIGURE 8.3

Until 1891 the German Empire's railways used five different operating times, the local times of the cities shown on the map. Just before his death Count von Moltke urged the adoption of a single time throughout the country, the one defined by the meridian (dashed line) one hour faster than Greenwich meridian time—today's Central European Time.

troops and the train crews being called up depend only on the clock time at their respective lodgings in their native land. The same condition holds with the timetables mailed to the railway administrations. At the present time, however, the North German railway administration uses only Berlin time; thus all tables and lists must be reworked into Berlin time. This repeated rewriting readily becomes a source of error, errors which can result in very serious consequences. This cumbersome process complicates the scheduling process enormously when tie-ups or accidents make immediate changes necessary.[10]

Noting the great advantage that would accrue from adopting a standard time for all German railways, von Moltke identified the fifteenth meridian east of Greenwich as the one best suited for the empire. Local time at the eastern border would differ from it by thirty-one minutes, and by thirty-six minutes at the western frontier with France. These differences were less than those to which Americans had become accustomed, and the selection of the Greenwich-linked meridian would mean even smaller differences for the south German states. But the great strategist also warned that a standard time for the railways alone would not eliminate all the disadvantages he had just cited. That would happen only if all local times throughout the German Empire were abolished and the proposed Greenwich-linked uniform time was adopted for all purposes.

As already noted, several groups—astronomers and allied specialists, factory owners, and representatives of agricultural interests—had all voiced strong objections to this proposal. Consequently, the rest of von Moltke's speech was devoted to refuting their arguments.

Focusing on "the savants of our observatories," von Moltke noted that astronomers and other scientists were not content with a uniform German time, nor with a Central European time; instead, they wanted a world time, a universal time. To desire it for their particular purposes was certainly within their rights, he acknowledged, but the time along the Greenwich meridian itself could not be introduced into the daily life of Germans; local times would have to be continued. Moreover, insofar as the railway enterprise was concerned, railway engineers were opposed to the use of a universal time. Conceding that astronomers recognized the need for railways to have a common standard of time, he derided their view that because travelers represented only a small fraction of the population, standard time should be used only for internal operations and not transferred to the public sector, pointing out that astronomers, geodesists, and meteorologists represented an even smaller fraction. And to opponents' argument that science required investigations and observations at specific locations and specific times, he answered that the corresponding local times could be determined "once and for all in the quiet of the study room."

Von Moltke was similarly dismissive of the arguments put forward by industrialists and farmers: that the shift in time would change the clock times of sunrises. Simply shift the opening times for the factories, he retorted, adding that farm workers routinely follow the sun and not their clocks. Further, if the courthouse bell strikes fifteen minutes early at the start of the day, workers will arrive fifteen minutes early—but they will also leave fifteen minutes early at the end of the day, so the duration of work will remain the same. Then, after refuting arguments related to mean time and apparent time, he ended his remarks by declaring:

Gentlemen, we cannot by a vote or a majority resolution decide on an arrangement which can be adopted only after negotiations with the Bundesrat; and perhaps also later via international agreements. But I believe that these negotiations will be facilitated if the Reichstag expresses its sympathy for a principle [uniform time for railways and telegraphs and some public activities] which has already been adopted, without substantial disruptions, by America, Sweden, Denmark, Switzerland, and South Germany.[11]

Six weeks later, on 24 April 1891, von Moltke died. Consequently, his last speech at the Reichstag, these forceful remarks on the subject of railway and public time, received worldwide circulation.

As the great strategist had stated, the advantages of using a single standard of time for the *internal* operations of Germany's railway system were obvious. Thus, on 1 June 1891, Prussian railway officials replaced Berlin time with Central European Time—Greenwich time plus one hour—a modest shift of slightly more than six minutes; public time-tables continued to be calculated in terms of local time.[12] However, one commentator predicted that as a result of von Moltke's remarks, the now-near-certain replacement of local time in the *external* service of Prussian railways would probably carry with it the adoption of the new time for civil affairs.[13]

Austria-Hungary had already agreed in principle to the new railway time, and the Prussian field marshal's remarks regarding its advantages with respect to mobilization undoubtedly resonated with government officials there. The empire's two railway operating times—Prague and Budapest—were replaced by Greenwich-plus-one-hour, a shift of two minutes and sixteen minutes, respectively. This change was adopted not only for internal use, but also for the railways' external services, including relations with the general public.[14] Not surprisingly, many Austrian and Hungarian cities quickly adopted the new time. However, Vienna chose not to do so. One commentator linked Vienna's refusal to the opposition voiced by the director of the Vienna Observatory, even though other Austrian astronomers, Robert Schram especially, were actively supporting the new system of uniform time.[15]

On the same day that Austrian and Hungarian railways adopted the new time, Romania replaced Bucharest time on its rail lines with the time of the third zone (*troisième fuseaux horaire*): Greenwich-plus-two hours, a shift of sixteen minutes. Many of the other Balkan countries soon followed Romania's lead.

Time unification continued. In 1891 the south German states (Bavaria, Württemberg, and Baden) announced that, not only for their railways' interior and exterior uses but also for their postal and telegraphic services, the change to Greenwich-plus-one-hour (Central European Time) would begin on 1 April 1892. The north German states later announced that they would adopt the new time on 1 April 1893. When that day arrived, the Imperial German government adopted the time for all state purposes, thus making the time of the meridian 15° east of Greenwich the empire's legal time. A similar bill was brought before the Austrian legislature.[16]

Over the next fifteen months other European countries—Belgium, Denmark, Holland, Italy, and Switzerland—adopted one or the other of the Greenwich-linked times. Progress toward time uniformity in Europe had been rapid. Not seven years had passed since astronomer Robert Schram had first written on the subject. He, the various railway directors, and von Moltke, buttressed by the successful switch in time in North America, had done their work well.

Although the actions taken in Europe to foster a uniform time system for the railways and the public are significant and certainly worthy of note, just as fascinating are the actions of two nonconforming nations: France and Great Britain.

Events in France commenced in January 1888 when M. F. S. Carnot, president of the republic, announced the formation of a commission to study the question of a single time—a legal or civil time—for the country. The commission, composed of members of the Bureau des Longitudes, submitted its report in June, with the recommendation that Paris time be adopted throughout France, Algeria, and Tunisia.[17] After obtaining agreement from all interested departments of government, President Carnot requested that the minister of public instruction and fine arts bring the matter before Parliament, and in November a draft bill was introduced in the Chamber of Deputies. The bill, presented in the name of the president of France and sponsored by the heads of ten ministries, proposed that Paris time be made the legal time for France and Algeria.[18]

The bill had been a long time in coming. In a summary of actions and its guiding role, the Bureau des Longitudes linked the subject to remarks made thirty-three years before by its current president, astronomer Hervé Faye, when he remarked that the adoption of the time of Paris for all of France was the inevitable consequence of its use as the operating time for the country's railroads.[19]

Left unanswered in the bureau's summary were two key questions: Why had it taken over thirty years to begin the formal process of adopting a national time? And why do so now? Undoubtedly current events, including the lack of French success at both the Rome conference of the IGA and the International Meridian Conference in Washington, were the catalyst.

France's time in 1888 was somewhat muddled—not unlike the situation in North America prior to November 1883. A contemporary article noted that three (actually four) systems were in use: (a) "local time," tied to place and varying with longitude; (b) "Paris time," or mean solar time along the meridian of the Paris Observatory; (c) "railway time," usually called *heure de rail* but in this article called *fautive d'heure de Paris*, or "defective Paris time"; and (d) "apparent solar time," of scant interest here.[20]

France's telegraph services used Paris time, transmitted every morning from the capital. So did most municipalities, which took the daily signal to regulate public clocks, adding or subtracting a constant amount to create local time for their inhabitants. However, like many towns in the United States prior to the adoption of Standard Railway Time, some French towns were using Paris time in place of local time. This practice was not sanctioned and obviously led to inconsistencies. What time was a particular municipality keeping? The answer was unanswerable without a direct, and often frustrating, inquiry.

Railway time in France was not fragmented to the same degree it then was in imperial Germany, for the railway companies had all adopted Paris time for their services—but only in principle. While exterior clocks at the stations displayed Paris time, interior timekeepers and rail operations were always five minutes behind this time—a situation unique to French railways. In effect France's trains were running not on Paris time, but on that of Rouen.

This fixed difference in clock times had been initiated soon after the dawn of French railroading, half a century before. Worried that travelers would be tardy, operating officials decided to provide a margin for them by setting all interior clocks five minutes earlier than the exterior ones.[21] As one critic of this difference observed, most of these early passengers actually arrived early on the platforms. Further, he noted, frequent travelers soon learned of the difference between the two clock times and counted on the "extra" five minutes. Certainly by 1888 there was no valid reason to continue the practice of having railway clocks at the same station display different times.[22]

Even though the Bureau des Longitudes was anxious to implement the new time, and even though it carried a weighty provenance, the Chamber of Deputies put the bill aside. Evidentially the long-held desire of a government bureau for uniform time throughout France was of little concern to the country's legislators.

Fifteen months passed before the Chamber of Deputies agreed to consider the issue of a national time again. As we have seen, by now (March 1890) Germany, Austria-Hungary, and Belgium were busily considering a Greenwich-linked time for their railways and public activities. Yet all this foreign activity had no impact at all on France. The bill before the chamber was just the same as before: to make Paris time the legal time for France and Algeria. Now, however, the climate in France had begun to change, for the country's savants were seeing articles in their journals questioning the lack of detail in the draft French legislation and urging consideration of expanding the system of time zones linked to Greenwich.[23]

This reconsidered bill might still have languished had it not been for a declaration of urgency issued by the Bureau des Longitudes–staffed commission and backed by the government. Consequently, in early December the Chamber of Deputies deliberated on the bill, passing it without dissent.

Action now shifted to the Senate. The prestige of the Bureau des Longitudes was tied inextricably to the bill; its defeat would be a major embarrassment for the institution, as well as for the government. To make certain that the strongest case would be made in favor of the bill, President Carnot appointed bureau president Faye a government commissioner to assist the minister of public instruction and fine arts during the Senate's deliberations.[24]

Discussion of the bill took place in mid-February 1891. Faye's remarks, subsequently described as an intermingling of patriotism with the question of uniform time, were calculated to deflect any consideration of alternate proposals. He began by saying that for France, the issue was an extremely simple one: with the exception of one eminent person, no one was demanding the adoption of "the English time"—a most unusual appellation and one certain to awaken negative feelings in the assembled senators. Asked by them to identify the dissenter, Faye named de Nordling, immediately dismissing de Nordling's views by claiming that the distinguished French and Austrian railway engineer and iron bridge designer "was brought to this question by very lofty scientific considerations," considerations that Faye did not feel bound to examine. "I do not believe that within these walls one can treat questions which may not be solved in a hundred years—like those of universal time," he remarked. And thus the "lofty scientific consideration" of railway station clocks showing different times was buried, never to be discussed. Faye even stated that the time displayed by clocks at the railway stations was already Paris time and, with passage of the bill, the entire country's municipal timekeepers would thereby display railroad time.

Faye's carefully worded statement regarding France's railroad time reflected a major concern: that many French cities and towns were using their own local times, and the scientist wanted all of them to adopt a single time standard. While agreeing that there would be a distinct altering of time in cities and towns far removed from the capital, Faye nevertheless argued that Paris's central location would minimize this shift's impact on the general public. His extreme cases were the city of Brest, whose local time was twenty-seven minutes slow on Paris; Nice, its time twenty minutes in advance of Paris; and Bastia, on the island of

Corsica, twenty-eight minutes in advance of the capital. Thus even the greatest shift from local time was still under thirty minutes. To demonstrate that the magnitude of such shifts would be tolerated by the public, Faye cited Great Britain, Sweden, and the United States, whose populations had already accepted comparable changes in their public times.

The proposed law was not coercive, Faye insisted. It "penetrates neither into the homes nor into the pocket of individuals for the sake of instructing what their watch says; everyone is free to set his own [time-keeper] as it suits him," for the language "stipulates only that all city clocks will display the same time." One can only conclude that the president of the Bureau des Longitudes included this disingenuous comment to divert the senators' attention. For in actuality, an individual would undoubtedly—and of necessity—set his watch and the clock(s) in his home to Paris time.

Returning again and again to the simplicity of the change, to its minor impact on the general public, and to the great benefits that would accrue once France's legal time and its railway time were the same, Faye ended his speech before the senate by remarking:

Gentlemen, I believe that no bill arises under simpler and more favorable circumstances. Its provisions will please all and constrain none. Most certainly the moment has arrived to regularize the current situation, to make the thousand small obstacles about which everyone complains disappear, and to return to regularity.[25]

The senators applauded and unanimously approved the bill. On 14 March 1891, "Legal time in France and in Algeria is the mean time of Paris" became law.[26] This national time applied to all government activities, to postal and telegraphic services, and to all railways.

Ironically, the five-minute difference between the interior and exterior clocks at the country's railway stations continued, for the railway companies were still in private hands and chose not to alter their operating times.[27] Although this was a minor remnant of inconsistency when compared against the virtually complete transformation of France's public time into a legally sanctioned, uniform one, the dynamics of change in neighboring European countries soon raised a fundamental question regarding the path the republic had taken. France, famous for its rationality and for its leadership role in gaining worldwide uniformity in weights and measures, was now out of step with the rest of the continent.[28]

It did not take very long, as such things go, for the resulting ensemble of issues to be raised in the country's legislature. In October 1896 a bill was introduced in the Chamber of Deputies that, if enacted, would change France's initial meridian from the Paris Observatory to that of

Greenwich, redefine the legal time of both France and Algeria as the mean time of Greenwich and also link the legal times of France's colonies to the world's expanding system of time zones (*fuseaux horaires*), and eliminate once and for all the distinction between the exterior times and interior times at railway stations.[29]

Although a few in France were in favor of the proposed change in the country's initial meridian, many others were not.[30] Without doubt, the long-standing opposition to Greenwich as the country's prime meridian by those in various government departments would have prevented the bill from becoming law. Not at all surprising then, that before this revolutionary bill had moved very far in the chamber, a substitute proposal, a *contre-project*, was put forth. Deftly circumventing the naming of the meridian, its thrust was simple: Only France's legal time had to be modified in order to gain the benefits of consistency with neighboring countries' zone times. The proposed substitute read, "The legal time in France and Algeria is the mean time of Paris, retarded by 9 minutes, 21 seconds"—this time interval the equivalent of the longitude difference between Paris and Greenwich.[31]

Introduced in March 1897, the bill, which was sent to committee and subsequently reported favorably, came up for discussion in the chamber late the following February. Accompanying it was a declaration of urgency. The bill's single article was adopted without opposition, and a few days later the proposed law was forwarded to the senate, which referred it to its own committee (see Figure 8.4).

Twelve and a half years passed before the committee placed its report before the senate, noting the extraordinary delay:

[The committee in 1898] found itself in the presence of discordant government opinions. While the Ministries of Trade, Industry, Posts and Telegraphs, and Public Works wholeheartedly accepted the proposition, the Ministries of Public Instruction and the Navy were formally opposed. Given these divergences, the Committee judged that it was not permitted to present its Report so long as the Government itself did not seem to have deliberated on the question and furnished the Committee a definite opinion.[32]

Although this explanation tends to stretch the bounds of credulity— would not a deputy or senator have raised the issue of a required but overdue report at least once during the prior decade?—it must be noted that the French government's formal approval of the proposal, without reservations, did not become available until 20 July 1910, just four months prior to the senate committee's submission.[33]

In any event, late the following January the senate began its deliberations.[34] Charles Lallemand, chief engineer of mines, whose proselytizing

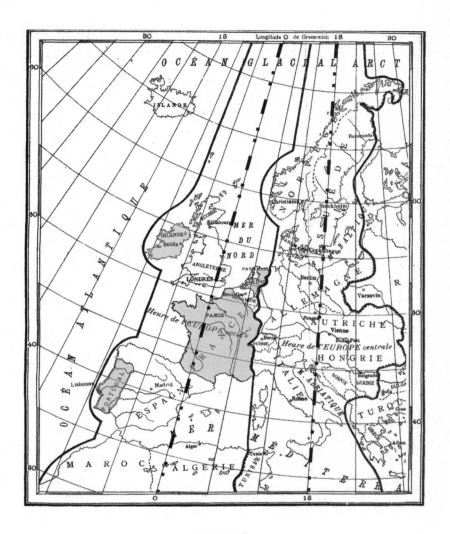

FIGURE 8.4

The adoption of zone time in Europe was rapid. By 1905 only three countries and the United Kingdom's Ireland (shaded areas) were observing neither Greenwich time nor Greenwich-plus-one-hour. The base map is taken from a more extended one in De Busschere, "L'unification des heures et son application en Belgique," facing 300. The boundaries and the time defining meridians of the two European zones shown have been enhanced.

had done so much to arrive at this juncture, was appointed government commissioner to assist the ministers of public works and posts and telegraphs.[35]

The discussions in the senate have been described as "lively." Although the senate report on the bill supported the government's declaration of urgency, the request was strongly opposed by several senators and dismissed. Indeed, one senator remarked that having exhibited patience in the matter for the past thirteen and a half years, it was ridiculous for the government to come before the senate and request such a designation.[36]

So the debate began. One opponent complained that the proposed nine-minute-and-twenty-one-second shift in time would increase the burden on those living east of the Paris meridian, who had already endured a sacrifice caused by the 1891 adoption of national time. His argument centered around the earlier clock times of sunset when local times were dropped in favor of Paris time. He claimed that in cities such as Nancy (sunsets fifteen minutes earlier by the clock) and Nice (sunsets nineteen minutes earlier), the result had been additional costs for artificial illumination, costs that would certainly increase were the time shifted nine minutes more. Moreover, those workplaces closely linked to daylight would suffer even more, particularly in the winter, if an additional shift were enacted. The senator reminded his colleagues that guided by the same considerations of light and dark, the English legislators were proposing to advance clocks by one hour in the summer. The government's response to this complaint—that while some French were penalized, others found advantage in the shift of legal times—was certainly not a very satisfactory one.[37]

Another opponent's prediction—that after the change to Greenwich time would come the abandonment of the Paris meridian—was publicly refuted. Both Lallemand and the bill's author insisted that not only did the proposed law not even touch upon sensitive national sentiments, it did not presume adoption of the Greenwich meridian. Indeed, the meridian of Paris would continue to be available for the requirements and products of French astronomy, French cartography, and the French navy—a position that was soon to change.[38]

After rejecting an amendment calling for the simultaneous use of Paris time and Greenwich time—which would have given France a dual time system—the senate adopted the proposed law as written, thus concluding the first day of debate on the bill.

During further deliberations held two weeks later, one opponent proposed major changes in the bill's language. His complaint was that the current text was inconvenient, particularly from the point of view of hydrography, and he proposed instead a very detailed wording that encompassed radio time signals based on Paris time transmitted from the military's Eiffel Tower station, the display of Greenwich time on the interior clocks of railways and Paris time on the exterior ones, and the regulation of marine chronometers based on the Paris meridian. As before, the wording meant a dual time system; in addition, the proposed language touched on the question of initial meridians. Again the government commissioner stated that this latter question had been set aside. A second vote on the proposed bill was taken, and a large majority voted in favor of it. The debate was over.

On 9 March 1911 the words that had been introduced over thirteen years before became law. All understood that the law represented adoption of Greenwich time as the country's legal time, even though the initial meridian's location was not mentioned. In addition, the French protectorate of Tunisia agreed to adopt Central European Time, and the long-standing difference between exterior and interior times at French railways stations ended. After years of debate and unyielding positions, France had embraced the expanding world system of time zones.[39] However, the Paris Observatory meridian continued as the nation's primary one for technical activities.

We turn now to the second nonconforming European country, the United Kingdom of Great Britain and Ireland. In the summer of 1880 the government decided to employ a defined legal time in its affairs. This action was ostensibly prompted by a letter in the *Times* from a court clerk pointing to ambiguities in the opening hours of polling booths and asking Parliament to resolve the issue.

It was certainly reasonable to consider what steps to take. Starting in 1840 the government had deliberately steered the kingdom's British railways to Greenwich time, which most had eventually adopted for both internal operations and public schedules. On the other hand, from 1858 on, the law courts had dictated use of a community's local time in business before judicial officials.[40] Despite this dichotomy, however, it was hard to discern any resulting difficulties inherent in having two times within the public sector, even by 1880, after over a generation of use.

In any event, the purpose of the statute was "to remove doubts as to the meaning of Expressions relative to Time occurring in Acts of Parliament, deeds and other legal instruments," with the language in the Definition of Time Act specifying that the time referred to shall "be held in the case of Great Britain to be Greenwich mean time, and in the case of Ireland, Dublin mean time."[41] And thus a *legal* dual time system—whose times differed by twenty-five minutes and twenty-two seconds—was inaugurated just as British specialists were starting to examine the issues associated with wider uses of the Greenwich meridian (see Figure 8.5).[42]

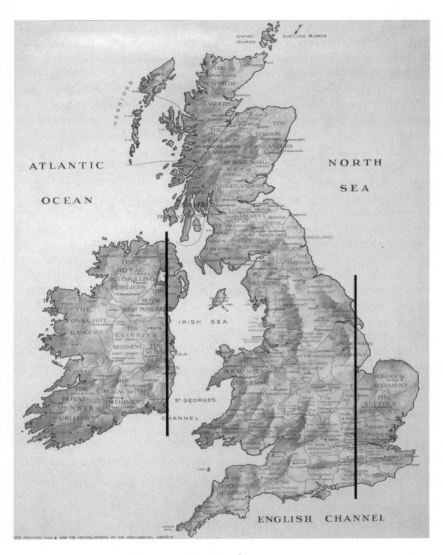

FIGURE 8.5

Between 1880 and 1916 the United Kingdom of Great Britain and Ireland had two legal times: Dublin time and Greenwich time, the former twenty-five minutes and twenty-two seconds slower than the time at the Greenwich meridian. The time meridians are the two solid lines. The base map is from a World War I poster; Library of Congress, LC-USZC4-11295.

As we saw in earlier chapters, since 1885 the government's policy and technical branches had been linked via an advisory committee created to consider the recommendations of the International Meridian Conference. Over the years, this committee's responsibilities in matters of time were broadened. Ireland's separate legal time was apparently never considered by this group of distinguished scientists, however, even though criticism of this difference as a detriment in advancing uniformity appeared regularly in the technical press as more and more continental countries began embracing a uniform time system.[43] Instead, after first focusing on altering the astronomical day and proselytizing in Great Britain for the twenty-four-hour notation, the committee began urging the adoption of Greenwich-linked time for public purposes in the British Empire's colonies. This was a path already suggested by the astronomer royal, who felt "a system of time based on hourly meridians . . . would be a convenient stepping stone to [universal time]."[44]

Astronomer Royal Christie's views regarding astronomical and civil times were in sharp contrast to those of his predecessor, Sir George Airy. Indeed, as one of the invited participants to the IGA conference in Rome, he had advocated the Greenwich meridian and associated time changes as early as 1883. Subsequently taking a leading role on the committee advising the lords of the Committee of Council on Education, the astronomer royal solicited support for change from the board of visitors to the Royal Observatory at Greenwich, whose annual report went to the Admiralty.

Christie also made his views known publicly. At the Royal Institution's weekly evening meeting of 16 March 1886, he delivered a lecture on universal time, repeating his conviction that hourly meridians based on Greenwich were only a precursor to the worldwide use of a single time.[45] He did concede that the use of hourly meridians might well work in relatively small countries—England, France, Austria, Italy, and the like—and that others particularly situated geographically—Switzerland, Greece, and New Zealand—might choose as their national (legal) time one an "exact number of half-hours from Greenwich."

But, the astronomer royal argued, countries having a great extent—the United States and Russia were his examples—would find the shifting of hours within their territorial boundaries so burdensome that eventually they would opt for a single time meridian, thus moving closer to a single universal time. Clearly he did not appreciate how cheerfully American railways and cities embraced the country's four time zones, or he might not have said, "When universal or world time is used for railways and telegraphs, it seems not unlikely that the public may find it more convenient to adopt it for *all* purposes" (emphasis added)."[46]

Christie's lecture undoubtedly led the committee advising the lords of the Committee of Council on Education to include the subject of hour meridians in time-defining suggestions for India and the colonies; however, this particular issue was so awkwardly described in the government's correspondence that it tended to be rejected out of hand.[47]

It was not until early 1890 that the subject of zone time was addressed again. This time the committee had before it the April article in the *Observatory* by Robert Schram detailing the imminent success of Greenwich-linked national times in Central Europe. Also before it was Sandford Fleming's "Movement for Reforming Time on a Scientific Basis," a lengthy memorandum sent by the governor-general of Canada to the Colonial Office and forwarded to the Science and Art Department.[48]

Fleming was still focusing on universal time (and on the twenty-four-hour notation); however, he was now advocating the reckoning of time in terms of hour meridians, or, as it was starting to be called, "hour zone time." He termed the alternate system "an easy transition from old to new ideas," and his memorandum included a table showing the "proper" zone times for the various colonies and for foreign countries as well as a world map delineating them. In July the advisory committee's recommendation that "Mr. Sandford Fleming's memorandum be forwarded to the Governments of all the Colonies for their consideration, with a view to the adoption of the hour zone system in reckoning time generally," was sent to the Colonial Office and to the India Office. Accompanying it was the committee members' "concurrence in Mr. Sandford Fleming's views as to the advantages which would result from this reform, and the ease with which it could be carried out."

Accordingly, in November 1890, the Colonial Office issued a circular despatch addressed to the governments of the empire's colonies.[49] In contrast to its earlier attempt, success came in February 1892, when the Cape Colony and the Orange Free State adopted a meridian 22°30'E, or one and a half hours in advance of Greenwich. These actions represented the first concrete result of nearly seven years of proselytizing efforts by the British government's advisory committee.[50] The fact of the imperial government's expressed interest, coupled with zone time's great success in North America and Europe, eventually led to its adoption by Australia (in 1895) and India (in 1905).[51]

Thus, by the first years of the new century, it was evident that zone time, civil times having exact hours (and half-hours) from the Greenwich meridian, eventually would be the world's time.[52] Left unresolved, however, was the question of adopting universal time.

CHAPTER 9

The French Take the Lead

[T]here are rare cases where . . . it might be advantageous to "shout"
the message, spreading it broadcast to receivers in all directions.
—*The Electrician*, 14 October 1898

By adhering . . . to the system of time zones, France has made one
of the last obstacles to *the unification of time* vanish.
—Charles Lallemand, 17 October 1912

It is easy to shake one's head and smile at how long it took the Republic
of France to adopt what had become the norm for civil timekeeping on
both the European and North American continents. Words like "chau-
vinism," "pride," and "stubbornness" come to mind. On the other hand,
bottling up a bill in committee is not unknown in other countries when
strong opposition arises. Indeed, opposition was still strong when the bill
finally came before the French senate a second time.

More significant than the reasons for inaction is the fact that the
French government did ultimately consider and adopt Greenwich time
legislation. Undoubtedly the growth of wireless telegraphy provides part
of the answer; the desire to create a leadership role for France appears to
supply the rest. Between 1910 and 1917 French actions resolved almost
all of the intractable issues associated with the Greenwich meridian and
uniform time.

In the 1890s, the stunning advances in transmitting information
longer and longer distances—across bodies of water, from the shore
to moving ships, between ships at sea, and between locales having neither telegraph nor telephone lines—were lost on no one, least of all the
military. Wireless communications equipment was purchased by Great
Britain in 1898 to use in the Boer War. That same year the Italian government adopted the Marconi Company's wireless system; and in July

1900 the British Admiralty ordered Marconi equipment for twenty-six naval ships and six coastal stations.

Commercial marine communications followed on the heels of these early acquisitions. By 1907 all large transatlantic passenger ships carried equipment able to transmit Morse-coded messages to, and receive them from, more than two dozen shore stations. Meanwhile German, American, and French scientists and engineers, some under contract to their respective militaries, were hard at work on new designs to circumvent the Marconi patents.[1]

The advantages of the new wireless technology led to ideas for its application in other areas, including time. As early as 1898 the Irish telescope maker and scientific-apparatus designer Sir Howard Grubb proposed that city clocks be regulated automatically by electromagnetic waves transmitted from a central location, a concept seconded a few years later by the astronomer Guillaume Bigourdan, a member of the Bureau des Longitudes. In 1903 a French naval engineer suggested transmitting time signals several times a day from shore stations for the regulation of shipboard timekeepers, the signals to be special ones—repeated once per second—so that they could be recognized even in the presence of strong interference generated by other transmitting stations. Then, in September 1904, the German astronomer and geodesist Theodore Albrecht, at the Geodetic Institute in Potsdam, determined a longitude difference between two sites 33 km apart, using the central transmitter of the Wireless Telegraph Company (Gesellschaft für drahtlose Telegraphie) to exchange signals. Albrecht announced plans for making wireless longitude-difference observations between Berlin and Karlskrona, on Sweden's south coast (450 km), and between Berlin and Vienna (550 km).[2]

Time signals broadcast via wireless were inaugurated by the United States in January 1905, shortly after the U.S. Hydrographic Office published a *Special Notice to Mariners* that included the statement, "Arrangements for a time signal service by wireless are now being made."[3] The second regular time service was established by the Marconi Wireless Telegraph Company of Canada at its Camperdown shore station, at the entrance to Halifax Harbour, Nova Scotia. Beginning in May 1907 automatic signals from the Observatory in St. John, New Brunswick, were telegraphed to this facility and the resulting wireless signals transmitted daily at 10 A.M. Atlantic Standard Time.[4]

In 1910 two high-power fixed installations began transmitting time. One was located at Norddeich, on Germany's North Sea coast, and the other was in Paris, the antenna there supported by the Eiffel Tower (see Figure 9.1). Starting in May, time signals telegraphed from the respective observatories were now distributed on a regular basis to much of Europe and broad sections of the Atlantic Ocean.

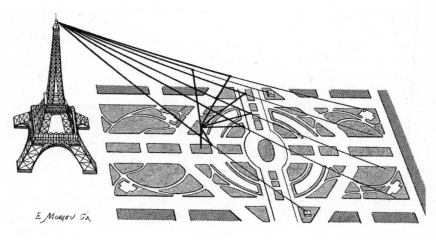

E. MORIEU SK

FIGURE 9.1

*Beginning in 1910, time signals were telegraphed from the Paris Observatory
to the army's wireless telegraph station for distribution to ships at sea and
for measuring longitude differences. The high-power transmitting facilities,
located underground in the Champ-de-Mars, included the long-wave antenna
attached to the Eiffel Tower. From* L'Illustration *132 (8 August 1908): 98.*

Although the signals from three of these stations were transmitted at
specified exact hours of Greenwich time, the French station's transmissions were in terms of Paris mean time. Not surprisingly, each station
used a different scheme for sending its own time signals.[5]

The inauguration of wireless time signals in France was the result of
discussions and planning by various members of the Bureau des Longitudes, whose initial ideas were focused on the critical need for ships at
sea to determine their longitude.[6] Among those proposing solutions was
Bouquet de la Grye, longtime advocate of the Paris Observatory meridian. His March 1908 plan included the use of an antenna at the Eiffel
Tower (300 m high), or one sited at the highest point (3710 m above sea
level) on the island of Tenerife in the Canaries so that a periodic time signal could be transmitted worldwide. (Increasing the strength of the signal
was not considered a difficult problem.) Immediately the French Navy
engineer Émile Guyou announced that the Bureau des Longitudes had
been studying the subject since late December. The proposal before the
bureau noted the recently established time service at Halifax Harbour,
the possibility that an ensemble of carefully located radiotelegraph stations could provide coverage at all points on the world's oceans, and that
the linked issues of time-code format and the choice of a base meridian

were international questions. Included in the proposal was the recommendation that a commission of technical experts and mariners from
interested countries should be formed and that the bureau should take
steps to call such a meeting.[7]

On 13 May 1908 the Bureau des Longitudes voiced the desire that "a
time service be installed as soon as possible at the Eiffel Tower, by way
of a test, with the goal of assisting the determination of longitudes."[8]
To do so meant, among other tasks, constructing a direct link from the
Paris Observatory to the military radiotelegraph site at the Champs de
Mars, hard by the Eiffel Tower—a distance of approximately two miles.
Following negotiations with the minister of war, who agreed to provide
the necessary funds for the entire project, construction began.[9]

Early in the planning period, the French Army Major (later General)
Gustave Ferrié (see Figure 9.2), remembered today as one of the world's
radio pioneers, proposed the distribution of two different sorts of signals. Ocean navigators would receive daily time signals at a fixed hour,
exact to about one-half second, sufficient for their needs. However, for
determining the precise longitude difference between fixed points, the
transmitted signals would need to be exact to about one-hundredth of a
second. No problems were anticipated in meeting the first requirement,
but the latter required the most careful attention to all the many details
associated with determining the time, maintaining it, and transmitting
it via the new technology.[10]

According to the French histories, the transmission of Paris mean
time from the Eiffel Tower site commenced on 23 May 1910, and the
facility was formally dedicated on 22 June. Over the next year various
experiments were conducted, including a determination of the longitude

FIGURE 9.2

*Radio pioneer Major (later General)
Gustave Ferrié, August 1913. In
charge of the Eiffel Tower wireless
station, he was a member of the
French-American team determining
the Washington–Paris longitude
difference. USNO Library glass
negative g167.*

difference between Paris and Bizerte, on the coast of Tunisia, a distance of 1550 km.[11]

At this juncture, two years of planning and construction for transmitting time to ocean vessels were complete.[12] Suggestions that the Bureau des Longitudes consider the creation of an international commission of experts to consider the many issues associated with the wireless distribution of time had already been tendered. That time signals from other radiotelegraphic stations had been, and would continue to be, in terms of Greenwich time was apparent. Given this environment, it takes only a small jump into speculation to conclude that at some point in the process, changing France's legal time was considered by the government's technical experts, several of whom were already in favor of making this shift. One can even suppose that these discussions included a decision to wait until a significant and very visible technical accomplishment had been achieved—the success of the Eiffel Tower time station, say—before reinvigorating the legislative process. In any event, the government's official letter to the French senate carries a 20 July 1910 date and includes the statement that all opposition to the shift had vanished.[13]

As already noted, France's revised legal time became effective on 11 March 1911, eight months after the government announced its approval. Even before final passage, articles in the press were informing the public that "the mean time of Paris, retarded 9 minutes and 21 seconds," was actually *l'heure anglaise de Greenwich.* When the change came, French citizens did not protest.[14]

Signals from the Eiffel Tower transmitter shifted to Greenwich time on 1 July 1911. Charles Lallemand (see Figure 9.3) and others who were shepherding the legal time bill through the senate the previous November had insisted that the Paris Observatory meridian would be retained for scientific needs; however, French astronomers were even then considering dropping Paris time in favor of Greenwich. The previous April the Bureau des Longitudes had asked Benjamin Baillaud, director of the Paris Observatory, to organize a meeting of the directors responsible for the world's principal astronomical annuals along with those astronomers most interested in ephemerides of stars. Its purpose would be to find ways to increase the number of apparent places of stars in the yearly publications without increasing annual costs.[15]

In his invitation and in follow-up correspondence, Baillaud asked the would-be participants for suggestions. A common response was to cease the practice of identical calculations of particular tables by different groups, one respondent terming it "an enormous dissipation of scientific energy" and proposing that the work be shared. Another urged the adoption of a uniform meridian for the major parts of all annuals.[16]

Charles Lallemand, ca. 1910. The French engineer was a leader in his country's 1911 adoption of Greenwich time and recommended the formation of an international bureau of time. Courtesy Presses de l'École nationale des ponts et chaussées.

Accordingly, sharing the workload, exchanging the necessary base data, and using one and the same meridian for the entire collection of ephemerides were among the key issues for discussion when the International Congress on Astronomical Ephemerides held its first session in Paris on 23 October 1911.

At the first session, use of the Greenwich meridian was immediately agreed to. However, during the discussions one participant noted that unless French hydrographic charts were also altered to reflect the adoption of a new initial meridian, there would be difficulties for those navigators using the separate, naval-focused extract from the *Connaissance des temps*, especially after it had been revised to suit the astronomers' desires. Moreover, to execute such a change in the mariners' charts would take many years, cost a great deal, and cause difficulties for them during the transition period.

Although it was agreed that such a change was necessary, no solution could be addressed at this conference, for the astronomers had no authority to act outside their professional boundaries. Thus, in 1911, the issue of using the Greenwich meridian for French hydrographic products still remained an open one, hopefully to be resolved at some later date.[17]

After the congress had completed its week of sessions and voted on numerous resolutions, including the adoption of the Greenwich meridian in all astronomical ephemerides as soon as possible, the Bureau des Longitudes voiced its unanimous adherence to all the proposals adopted. It also agreed to recommend to the ministers of public instruction and the navy "the adoption, completely and without reserve, of the Greenwich meridian as the fundamental meridian in the *Connaissance des Temps*."[18] An era was rapidly coming to an end.

In the run-up to the soon-to-come 1912 conference that France hoped would solidify its leadership in matters of international time, one technical issue stands out: how accurately was the transmitted time being received? More than thirty years before, American astronomers had learned that their judgments regarding the quality of observatories' telegraphed time signals and their reception were wildly optimistic.[19] As a matter of fact, insofar as the various techniques employed to determine time at an astronomical observatory were concerned, little had changed in the ensuing decades. More often than not, specialists who promote new techniques and technology are overly optimistic.

A few months after the French and Germans initiated the distribution of time to mariners, Ferrié, who was in charge of the radiotelegraph station at the Eiffel Tower, described this important application at a January 1911 meeting of France's Society of Physics: "[F]or the regulation of chronometers the precision which one can obtain is certainly a half second approximately or a quarter of a second. . . . [The precision] can reach a tenth of a second for an experienced observer."[20] Evidentially Ferrié's statement was not based on actual data; within the year he would become aware of a more sobering, and apparently nonpublic, analysis.

That is not to say, however, that as-received accuracy was not a consideration early on. As a means of compensating for the inevitable uncertainties in the precise time of the Paris Observatory's master clock during long periods of cloudy weather, Bigourdan, president of the Bureau des Longitudes, asked his country's observatory directors to compare the time as determined locally by them with the Paris Observatory wireless signals from the Eiffel Tower station and to submit their results for study. By October the program was underway.

Observatories outside France were initiating similar comparison programs, but with the critical feature of comparing signals from *two* transmitting stations: Norddeich and the Eiffel Tower.[21] One of the earliest to do so was the Hamburg Observatory in Bergedorf—200 km from the German station and 730 km from the Paris one—which began acquiring time signals in August 1911. Slightly later comparisons document striking differences between the two transmitting stations (see Figure 9.4). Indeed, given the observing skills of these on-land specialists, the results

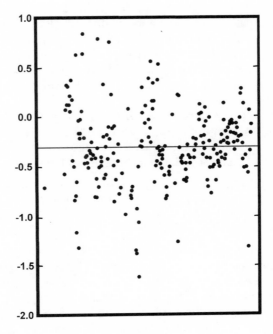

FIGURE 9.4

*Time differences, in seconds, between two wireless stations based on
measurements at the Hamburg Observatory, Bergedorf, during the year 1912.
The average difference (Norddeich – Paris) is shown by the horizontal line at
–0.31 sec. Data from* Jahresbericht der Hamburger Sternwarte in Bergedorf für
das Jahr 1912 *(Hamburg, 1913): 26–33.*

are more significant than a comment made later that year by Ferrié that
"according to the statements of ocean navigators, the time signals emitted by the Paris and Norddeich stations frequently differed by one and
two seconds."[22]

A major problem had been uncovered at Hamburg, one that would
not be addressed until after World War I, and that would require more
than two decades of intense effort among the world's geodesists and
observatory astronomers to overcome.[23] Moreover, it is somewhat
amusing to read in a recent history of this early era that, after enumerating a number of subtle effects contributing to these unexpected
differences, the author wrote matter-of-factly, "perhaps one cannot be
surprised that the signals were emitted with variations of the order of a
second compared to the time they were supposed to provide."[24] In fact,

given the contrasting statements of Ferrié, the early practitioners must have been astonished to find such large differences for the first time!

A second issue came to the fore during the run-up to the international conference on time: the shocking loss of 1,500 men, women, and children on 15 April 1912 when, on its maiden voyage, the White Star liner *Titanic* struck an iceberg and sank. Both the United States and Great Britain began official inquiries.

During the June sessions of the British inquiry a copy of a wireless telegram from the captain of the French steamship *La Touraine* to the captain of the *Titanic* was introduced (see Figure 9.5). Transmitted two days before the *Titanic* struck the iceberg, it read in part:

My position 7 P.M. Greenwich, latitude 49 28, longitude 26 28 w. Dense fog since this night. Crossed thick icefield, latitude 44 58, longitude 50 40, Paris. Saw another icefield and two icebergs latitude 45 20, longitude 45 09 Paris. Saw a derelict latitude 40 56, longitude 68 38, Paris.[25]

This mix of a Greenwich-based time and a Paris-based longitude did not go unnoticed, though any ocean navigator using Greenwich would have been able to adjust the French ship's longitudes with ease. The *Observatory*, however, targeted the use of the Paris meridian for longitude, voicing the hope that it was seeing a "transitional period . . . which will soon pass."

The French government, acting on the initiative of the Bureau des Longitudes, had invited delegates from foreign countries to participate in an international conference to be held in Paris "to study the means for realizing the practical unification of time" and also to discuss the formation of an international time service organization to lead in this endeavor.[26] From 15 to 23 October 1912, delegates from sixteen countries met to consider the issues associated with the unification of time. A few of the delegates had already participated in the June wireless telegraphy conference in London, at which recommendations were formulated regarding radio transmissions, including time signals from coastal stations and similar services, to help prevent disasters at sea.[27]

One of the American participants at the Paris conference, the Naval Observatory astronomer Asaph Hall, Jr., wrote in his confidential report that good reasons existed for calling an international conference: With several countries already transmitting time signals, there was a need to ensure that they were alike, insofar as possible, thereby "enabling a navigator in any part of the world to know what signals to look for." He also remarked that "a weighty reason with the French for calling the conference is that they wished to centralize this [unification] work in Paris," for, with a high-power station near Berlin under construction and another one in Belgium nearly operational, they "felt that in order

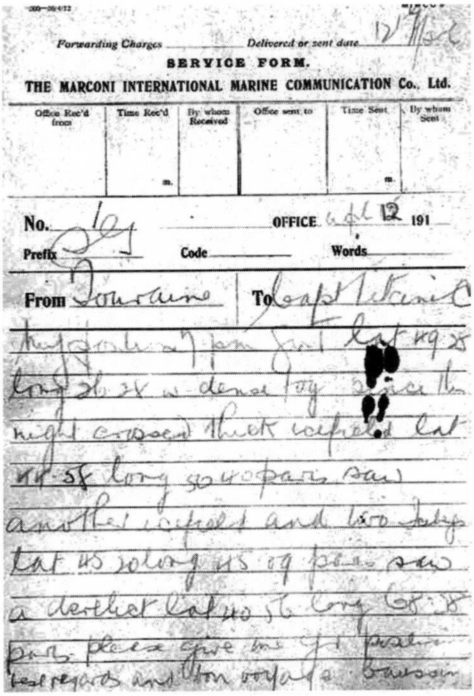

Forwarding Charges ______ Delivered or sent date 12/4/12

SERVICE FORM.

THE MARCONI INTERNATIONAL MARINE COMMUNICATION Co., Ltd.

Office Rec'd from	Time Rec'd	By whom Received	Office sent to	Time Sent	By whom Sent
	m.			m.	

No. 1/2 OFFICE April 12 191

Prefix ______ Code ______ Words ______

From Touraine To Capt Titanic

My position 7 pm GMT lat 49.28
long 26.28 w dense fog since this
night crossed thick icefield lat
44.58 long 50.40 paris saw
another icefield and two [illegible]
lat 45.20 long 45.09 paris saw
a derelict lat 40.56 long 68.38
paris please give me yr position
best regards and bon voyage Caussin

FIGURE 9.5

Telegram from La Touraine *to* Titanic *giving its time in terms of the Greenwich meridian and its longitude in terms of the Paris Observatory meridian. Reproduced with the permission of the Marconi Corporation, plc.*

to keep the preeminence of the Eiffel Tower signals they must secure something like an international sanction."[28]

During the week of meetings, various technical issues were discussed. In addition, Charles Lallemand presented, in the name of the Bureau des Longitudes, a "Plan for the Organization of an International Time Service." He recommended that it be separate from, but similar in international status to, the International Geodetic Association and the International Bureau of Weights and Measures. Such a service, consisting of a central bureau linked to a central transmitting station, would collaborate with the time-distributing observatories throughout the world. It would analyze the time transmissions from the other radio stations and then determine, as closely as possible, a unified time based on the Greenwich meridian.[29]

This proposal for an international bureau was accepted by the delegates, and steps were taken—initially just after the conclusion of the 1912 conference, and then in 1913 via the Second International Conference on Time. This technical conference was followed immediately by a diplomatic convention, at which it was agreed to establish an international association of time. The outbreak of World War I prevented the ratification of the convention. Nonetheless, the Greenwich meridian was now the unquestioned basis for both technical and public timekeeping.[30]

Also accepted in 1913 was the use of the Greenwich meridian on hydrographic charts issued by all European maritime nations. As noted above, those participating in the October 1911 International Congress on Astronomical Ephemerides had raised the issue of adopting the Greenwich meridian on France's hydrographic products so that the various astronomical tables, including those used by navigators, would all be in terms of that meridian. At the conference several French officials had stated that the time was not quite ripe to make such a change.

Undoubtedly the loss of the *Titanic* completely changed the environment. Sixteen months after the 1912 disaster a French ministerial decision was announced. It directed that "the longitudes shown on French nautical documents published after 1 January 1914 are [to be] reported in terms of the international meridian of Greenwich."[31] The directive meant that any ambiguity in describing position no longer existed. In the fall, after two months of discussions, the world's maritime nations signed the *Convention for the Safety of Life at Sea*. The abbreviations adopted for radiotelegraphic messages did not include a cipher specifying an initial meridian when longitude information was transmitted, for all understood that it was the Greenwich meridian.

With France the last European maritime nation to adopt the Greenwich meridian on ocean charts (see Table 9.1), the stage was set for further unification efforts.[32] Already France and other European countries had been urging the creation of maps for air navigation, of which Charles Lallemand was a prominent early advocate. His espousal of maps covering one degree of latitude by one degree of longitude, with Greenwich as the initial meridian, received worldwide attention.[33]

The outbreak of World War I in August 1914 halted international uniformity efforts. The protagonists focused their energies on harmonizing their own activities, with the Allies coping with nonuniformity among the large-scale French, Belgian, and British topographical maps—differing initial meridians, differing scales, differing projections, differing grids and sheet sizes—as they rushed to prepare maps needed for trench warfare.[34] When in 1916 the warring Allied and Central Powers' nations adopted different periods of daylight saving time, progress toward global

TABLE 9.1

*Initial meridians on hydrographic charts,
1914 (compare Table 2.1)*

Country	Meridian
Austria-Hungary	Greenwich
Belgium	Greenwich[1]
Denmark	Greenwich
France	Greenwich[2]
Germany	Greenwich
Great Britain & Ireland	Greenwich
Italy	Greenwich
Netherlands	Greenwich
Norway	Greenwich
Portugal	Greenwich[3]
Russia	Greenwich
Spain	Greenwich[4]
Sweden	Greenwich
United States	Greenwich

1. Belgium dropped the Paris meridian on its maritime charts around 1900.
2. France adopted the "International meridian" on 1 January 1914.
3. Portugal had adopted the Greenwich meridian by 1913.
4. Spain adopted the Greenwich meridian for its navy on 3 April 1907.

uniformity was seemingly dealt a second setback.[35] Rather unexpectedly, however, the adoption of advanced time by the Allies actually improved uniformity in two areas: time at sea and astronomical time. In both cases the French led the way.

Late in January 1917 Joseph Renaud, director of the Hydrographic Service of France's Ministry of the Navy (see Figure 9.6), wrote Rear Admiral John Parry, Great Britain's Hydrographer of the Navy. His subject was the time kept on board ships, specifically "the time upon which the internal arrangements of vessels is based and the time which is noted upon all documents kept on board."[36] Renaud was soliciting Parry's views prior to any official action. However, Parry was ill at the time, so Renaud's letter was acknowledged and answered as if it were an official inquiry to be handled by the responsible naval attaché. By the time the admiral returned to work and was able to apologize for the actions that had been taken in his absence, the French government was already poised to carry out its own decisions regarding timekeeping at sea.[37]

Without doubt the subject had been under review for some time, for in mid-January Renaud put the finishing touches to a detailed, carefully reasoned analysis titled "Time at Sea."[38] In it he introduced the issues by

FIGURE 9.6

Joseph Renaud, ca. 1917. The French hydrographer, a leader in unifying hydrographic information, proposed and put into practice the use of standard time zones at sea. Courtesy International Hydrographic Bureau, Monaco.

noting that prior to France's adoption of legal time (Paris time) in 1891, vessels at sea or in a harbor regulated the time displayed on their clock by apparent time. Since May 1891 the country's vessels had been obliged to use legal time when at French and Algerian ports and harbors, while they continued to use apparent time when at sea. When in 1911 legal time shifted to the time of the Western European zone (Greenwich Mean Time), ships at sea continued to use apparent time.

Renaud also described the process of changing the ship clock's display to make it conform to the apparent time as determined at midday, noting that the change was generally made at midnight, but more often when the vessel was changing her longitude rapidly. He commented on the subsequent need to adjust the crews' on-duty times (watches), and on how special rules had to be promulgated when in 1916 the legal time of France was advanced sixty minutes.[39]

Renaud then turned to the advantages associated with adopting zone time at sea. Noting that transmissions of wireless time signals were in terms of universal time (Greenwich Mean Time) and that meteorological signals were also transmitted at agreed-upon times, he reminded his readers that the *Titanic* disaster had led to the international agreement to report the location of derelicts and ice in the Atlantic sea lanes. These wireless reports were sent in code, with the associated times in terms of the international meridian (Greenwich Mean Time). Continuing to use apparent time on board ship, adjusted at midnight, meant that the calculation for the vessel's precise longitude might be in error, or at best ambiguous. Adopting zone time would eliminate such ambiguities.

Of course, adopting zone time on ships at sea would lead to complications arising from the need to add or subtract sixty minutes the moment a vessel passed from one zone to another—a very large shift compared with the usual changes in ship time under the current system and one that might need to be made at any time of the day or night. Such a change would strongly impact the length of a crew's watch. Renaud suggested that the time tables governing a ship's routine be controlled to reduce the impact of changing time by one hour, and in any event, "the inconveniences resulting from an alteration of 60 minutes . . . would not be greater than those which actually exist, when a vessel alters her clock from apparent to legal time on arriving in port or vice-versa."[40]

Renaud also commented on other methods that might be used for keeping time at sea, one being the use of universal time for events of interest to others and continuing with the use of apparent time on board ship. This would require two clocks, and thus a double system of timekeeping. As on land, such requirements were unsuitable. Renaud concluded:

[I]t may be said that the method of time keeping actually in use by vessels at sea, entails serious inconveniences which have been more especially felt since science has made increasing demands upon the degree of accuracy necessary in all forms of observations, and since wireless telegraphy has put vessels at sea into common communication with other vessels, and with the land.[41]

Thus a French government specialist had put forward a recommendation to have the system of time zones encompassing the world's land masses (*fuseaux horaires*) applied to the oceans.

After completing his analysis and writing to Parry for his views, Renaud presented a brief note to the Académie des Sciences. Then he submitted his analysis to the Bureau des Longitudes, along with a joint proposal with Lallemand that the bureau review the recommendations with a view to having zone time adopted by the French navy.

The Bureau des Longitudes unanimously approved the proposal and presented it to the minister of public instruction for transmittal to the minister of marine. In mid-March the navy minister accepted the proposal and directed that beginning on 25 March, time at sea on warships and merchant ships under French navy control would be in terms of zone time. Less than two months had passed since Renaud completed his analysis.[42]

All that remained for the French government to do was to place notices of the minister's decision in the *Journal officiel de la Republique*

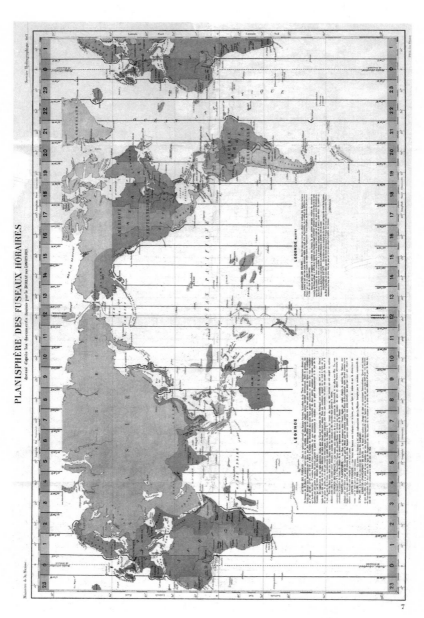

FIGURE 9.7

World time zone chart issued in June 1917, from a copy at the Library of Congress. D'après l'ancienne carte de SHOM "Planisphère des fuseaux horaires." Autorisation de reproduction no. 553/2005.

7

Française, define and publish the limits of each time zone, and prepare a chart showing the zones. (The "Planisphère des fuseaux horaires," the first-ever official chart of the world's time zones, appeared in June; see Figure 9.7.)[43] However, because of Parry's absence, in mid-March the British had taken no steps at all in the matter—much less considered the issues. Obviously, for any system of timekeeping at sea to come into general use, the world's largest sea power had either to accept it totally or propose modifications.

Accordingly, soon after his return from sick leave Parry proposed that the Board of Admiralty issue invitations to a "Conference on Time-Keeping at Sea." The conference, which opened on 21 June 1917 with Parry as chairman, included representatives from a host of government departments and scientific societies.[44] Renaud was specifically invited because of his leadership role in maritime timekeeping and his position as a world authority on the subject. As Parry observed in his opening remarks, "The adoption of a system of Time Zones in the French Navy was of such immense importance that it must inevitably, sooner or later, become a matter of international importance since every branch of Maritime science was involved."[45]

The committee met four times, completing its work on 3 July. As one might expect, having the benefit of copies of Renaud's analysis, the recently printed French chart, in color, of the world's time zones, and the actions taken subsequently by the French government, there was essentially complete agreement among the British representatives regarding the use of zone time by vessels of His Majesty's navy. Consequently, the recommendations in the report to the lords commissioners of the Admiralty included adopting the time zone boundaries as defined by the French, along with all their other specific recommendations, including those for exchanging signals and wireless messages. Indeed, the only major difference between the experts of the two countries was the notation for the time zones themselves. The French were numbering them from 0 to 23 eastward, starting with the Greenwich meridian zone; the committee recommended that, starting with the Greenwich-based zone as 0, zones west of Greenwich should be described as plus 1, plus 2, and so on, ending with plus 12 for that part of the zone lying east of the date line; and minus 1, minus 2, and so on for zones east of Greenwich, ending with minus 12 for that part of the zone lying west of the date line.[46]

The Admiralty committee was unstinting in its praise of Joseph Renaud, expressing the hope that

a suitable acknowledgment of the great indebtedness to Monsieur Renaud might be officially conveyed to the French Government . . . not only for his valuable assistance to the Conference, but also on behalf of seamen generally, for raising this most important question, the discussion of which may eventually lead to a general uniformity in time-keeping throughout the world.[47]

It is noteworthy that the Admiralty did not adopt its committee's recommendations until nearly two years later despite the level of agreement among the several departments of the British government and the interested scientific societies. Indeed, the Italian government agreed formally to adhere to the zone time system while conference members were still in the midst of their deliberations—and less than six months after it had first been apprised of the French efforts.[48]

Clearly, the issue of timekeeping at sea was not one of high priority for the British, because its warships had long been operating under Greenwich time throughout the so-called Home Water—a zone extending from 45°N to the northern limits of the European seas, and from the coast of Europe to 30°W. Consequently, there was no strong need to change.[49] Moreover, Germany's second unrestricted warfare campaign was still at its height, and undoubtedly all efforts were focused on measures to stem the enormous loss of shipping caused by the enemy's submarines. In addition, no charts of the kind already available to French vessels existed. Further, steps had to be taken to have Britain's merchant navy consider (and hopefully agree to) the change in timekeeping.[50]

One other issue remained. In what must have been a compromise position taken as a result of differing views—for already the French had scheduled this particular change for 1920—the Admiralty committee's eleventh and last recommendation concerned the reckoning of the astronomical day:

That from the point of view of seamen it would be of considerable advantage if a day commencing at o hours midnight were substituted for the astronomical day commencing at o hours in all nautical publications, and that the Royal Astronomical Society should be asked to ascertain the views of astronomers as to such an alteration, including the possible substitution of the civil for the astronomical day.[51]

The process of having astronomers consider such a change began in a strangely obtuse manner. Despite public mention of the Admiralty's conference on timekeeping appearing in *Comptes rendus*, including the recommendation for the substitution of civil time for astronomical time, the inaugurating letter from the astronomer royal, F. W. Dyson, and the Oxford astronomer H. H. Turner (who represented the Royal Astronomical Society at the conference) began with a reference to the expression-of-hope resolution adopted at the International Meridian Conference of 1884 and continued: "Our attention has lately been again drawn to the subject." Not a single word in the letter referred to the recent conference;

indeed, the only evidence proffered in favor of making such a change was an undated general note from the Board of Trade's representative there, who was not identified.

In their 17 July letter, which was published in the *Observatory*, Dyson and Turner expressed the view that it was desirable "to have general agreement among astronomers before such a change were made," but they emphasized, "[a]ny decision on the matter must rest with the authorities responsible for the national ephemerides"—which, in Great Britain's case, was of course the Admiralty. The two invited their colleagues' opinions on the matter, to be communicated to the Royal Astronomical Society.

A follow-up article in the September issue of the *Observatory* did nothing to clarify matters, for it was simply a translation of the late Henri Poincaré's 1895 report supporting the change in the astronomical day—important, of course, but from an era long gone. Only the last journal communication on the subject can be considered an attempt to address some of the current issues, but it, too, made no mention of strong Admiralty interest in the matter.[52]

Why did the two astronomers take such an indirect approach? Were they worried about naval secrecy? Were they being chauvinistic with regard to French leadership in maritime timekeeping operations? Had the fact that the first British public communication on the subject bore the signature of the astronomer royal given the country's astronomers an unmistakable signal that the government supported the change? Or, perhaps, was the report of the committee not yet available to the lords of the Admiralty for distribution, so that Dyson and Turner's July letter merely served to raise general awareness? In any event, the official request from the Admiralty to the Royal Astronomical Society was sent on 15 December 1917, after an informal inquiry in October.[53]

The following February a committee appointed by the Council of the Royal Astronomical Society, meeting to consider changes that would be required in the *Nautical Almanac*, determined these were few in number and could easily be made. In mid-March 1918, the society issued a circular letter soliciting the opinions of "Astronomical Societies, Superintendents of Almanacs, Directors of Observatories, and other representative astronomers in the allied and neutral countries as to the desirability of the proposed change [of the astronomical day]."[54] The Society's letter noted that France intended to make the change in 1920, a decision French specialists had taken in April 1917.

As the Royal Astronomical Society's inquiry got underway, other venues became aware of the London conference and its recommendations. In February the *Geographical Journal* published a long summary of the recommendations, noting that the conference's report had been presented to the lords of the Admiralty. The Marine Department of the Board of Trade prepared a "Memorandum on Time Keeping at Sea" and distributed it widely, together with a chart showing the proposed zones, to Britain's merchant fleet. Additional articles on the subject appeared in French journals, as well as two in *Nature* written by the secretary of the Royal Astronomical Society.[55]

Throughout this period, apparently no one remembered the massive effort to alter the astronomical day mounted by Sandford Fleming twenty-five years before. Moreover, none of the professional journals reminded readers that Dyson's predecessor, William Christie, had been a leader in the British government's attempts to gain approval for this change from the producers of almanacs.[56] So the same issues that had surfaced then had to be considered anew.

Many astronomers—the ones opposed to the change—expressed misgivings concerning the notion that the proposed alteration in the astronomical day would actually help seamen. As a consequence, the Royal Astronomical Society asked the Admiralty for further information to support the change. The response, printed in the *Monthly Notices*, probably did nothing to change the views of doubting astronomers. Nonetheless, its content certainly should have made anyone reading it aware that the Admiralty strongly favored the change.[57]

Over the next months the responses to the Royal Astronomical Society's inquiry trickled in.[58] Without any doubt, the most important one was the reply from the U.S. Naval Observatory, for the Great War had turned the "big four" nautical almanac producers into the "big three": France, Great Britain, and the United States. The USNO was in favor of the change, as of course was France. All that remained was "housekeeping": the preparation of a summary response that would support the Admiralty when it modified its product. And so on 21 December 1918 the Council of the Royal Astronomical Society reported in favor of the change, adding, "The Council consider that it is practical and desirable to introduce the change into the Nautical Almanac commencing with the year 1925."[59]

The Admiralty moved swiftly. On 17 January the superintendent of the Nautical Almanac Office, responding to an inquiry from his superiors, indicated the change could be made in the *Nautical Almanac* commencing with the 1925 edition. On 29 January he was directed to do so. In mid-February the Admiralty informed the Royal Astronomical Society of the directive from the lords commissioners of the Admiralty to its superintendent.[60]

With the alteration of the astronomical day finally out of the way, the Admiralty turned to the fundamental issue: timekeeping at sea. The decision of their lordships came early in April. Associated with it was a "World Time Zone" chart that depicted the extent of the zones. With the exception of the zone numbering for determining Greenwich time, the zones were identical to those shown on the earlier French chart.[61] The following year the U.S. Navy announced the use of zone time by its vessels, calling attention to its Hydrographic Office chart, "Time Zone Chart of the World," which followed the British zone numbering.[62]

With the publication of these official statements, the transformation to zone time by the "big three" was complete. Their actions undoubtedly provided the impetus for other countries to adopt zone time as their legal time on land. Thanks to the persistence of the French hydrographer Joseph Renaud, this giant step in the unification of time had taken only three years.[63]

PART III

Employing Clock Time as a Social Instrument (1883–1927)

CHAPTER 10

Advancing Sunset, Saving Daylight

I will not dispute with these people, that the ancients knew not
the sun would rise at certain hours; they possibly had, as we have,
almanacs that predicted it; but it does not follow thence, that they
knew *he gave light as soon as he rose.*
                                    —Benjamin Franklin, 26 April 1784

The seasonal meddling with the clocks cannot . . . be justified from
a scientific point of view. . . . [F]or a nation to do this by legislation
would be the height of folly.
                                    —*Nature*, 11 June 1908

The [Summer Time] experiment has been an unqualified success,
and, as happens in all similar cases, one is astonished that it was
never tried before.
                                    —"Summer Time," *Times* (London), 26 August 1916

While specialists were trying to extend uniformity, both in timekeeping
and the use of the Greenwich meridian, a completely unexpected approach to the use of time, time as a social instrument, came to the fore.
Though at first considered damaging to time uniformity, the concept
gained acceptance during World War I. Today, periodic time change is
perhaps the only disputed area of importance in the art of public timekeeping. How such alterations in civil time became part of our culture is
the subject of this and the following chapter.

Altering the time displayed on a clock or watch is a common enough
practice. Many people set their personal timepieces a little fast, hoping
to create a cushion that will help them to avoid being late for appointments. Others keep the correct time but set their bedside alarms to wake

up early. Only a very few are so obsessed with time that they would keep all their clocks in synchrony with a radio- or satellite-controlled watch—even resetting clocks in motel rooms and relatives' apartments so that they conform. For such human timekeepers, advancing official time and then retarding it periodically is a mixed blessing. While it allows them to indulge their time-tinkering obsession, it can become a burden, considering the vast number of time-controlled devices in today's homes: TVs, VCRs, DVDs, computers, thermostats, answering machines—not to speak of clocks in every room. Perhaps it would be better not to alter time displays but to change people's habits instead.

Altering Parisians' daily habits was precisely what the philosopher and man of letters Benjamin Franklin (1706–1790) had in mind when, in 1784, he sent an extremely witty letter to the *Journal de Paris*, a popular daily.[1] Actually Franklin intended this piece as a parody on the scientific method. His communication began with this report: Franklin the scientist was unexpectedly awakened by sunlight streaming into his bedroom six hours before his normal rising time of noon. He consulted an almanac for that day—20 March, the first day of spring—and found that the sun was scheduled to rise at 6:00 A.M.; moreover, not for six months would it rise any later. Franklin repeated his observations on each of the following three mornings, with the same, astonishing result: *"that he [the sun] gives light as soon as he rises."*

Armed with this "discovery," for which he asked no reward or pension, imbued with a love of economy, and convinced that "a discovery which can be applied to no use . . . is good for nothing," Franklin proceeded to offer his proposal. Judging that, like him, Parisian families rose at noon, he calculated the monetary savings that would accrue if all citizens instead rose with the sun during the spring and summer months and used its light for illumination instead of staying up long after dark and burning candles. The savings would be enormous (and unrealistic): 96,075,000 livres tournois—over $8 billion in U.S. dollars today.[2] Franklin concluded his one-sided cost-benefit analysis by asserting that his calculation was a conservative one, for it applied only to half the year, and "much may be saved in the other [half], though the days are shorter." Further, given the resulting unburned stocks of wax and tallow, candle prices would drop significantly in the following winter—and would continue to be cheaper "as long as the proposed reformation shall be supported."

As to the chances of realizing his proposal in practice, Franklin professed to hold a strong belief in the common sense of Parisians. After reading of his discovery of the light given by the sun as soon as it rose, most, he felt sure, would immediately start rising at dawn. However, knowing that a few of the city's inhabitants were without common sense, he proposed the use of coercive measures. They were (1) a heavy tax of one louis (more than $1,300 in today's dollars) on all windows having shutters to keep out the light of the sun; (2) the stationing of "guards in the shops of wax and tallow chandlers, and no family be permitted to be supplied with more than one pound of candles per week"; (3) the posting of guards "to stop all the coaches, &c. that would pass the streets after sunset, except those of physicians, surgeons, and midwives"; and (4) the ringing of church bells at sunrise. Moreover, if the sound of their peals proved insufficient, "let cannon be fired in every street, to waken the sluggards effectually, and make them open their eyes to see their true interest."

Franklin predicted that "all the difficulty will be in the first two or three days," having observed that "oblige a man to rise at four in the morning, and it is more than probable he will go willingly to bed at eight in the evening." And, after a night's sleep of eight hours, "he will rise more willingly at four in the morning following."

Franklin, who left Paris in 1785, made no attempt to lobby France's then-royal government for the decrees needed to change Parisians' habits. However, so successful was he at maintaining a strict pedagogical style in his straight-faced parody that some American and British writers have actually taken his proposal seriously and portrayed him as the father of today's daylight saving time. Although this multitalented American genius may well be the father of studies of the Gulf Stream and of lightning, he certainly did not create the back-and-forth time system in effect today.

Throughout the nineteenth century, the consumption of candles and tallow for illumination continued, joined by ever increasing quantities of coal-derived gas and electricity. The issue of time remained largely a matter for scientists or those responding to special needs by legislating a legal time for public activities.

Early in the twentieth century a true champion of sunlight arrived on the scene: William Willett (1856–1915; see Figure 10.1), a London builder of fine homes. Even though a proposal identical in concept to his had been put forward twelve years earlier, Willett is the father of today's system of advancing the clock periodically.[3]

According to the available biographies, Willett was an avid horseman and golfer.[4] During early morning rides from his home in suburban Chislehurst southeast of London, through St. Paul's Cray Common and on to Petts Wood, he discovered the influence of the sun. As he rode by the area's large homes, Willett observed that their window blinds were closed to keep out the morning light. From these observations came his essay *The Waste of Daylight*.[5]

FIGURE 10.1

William Willett, ca. 1910. In 1907 the London builder published The Waste of Daylight. *From it stemmed today's Daylight Saving Time and Summer Time. Image from 1914 edition, courtesy Bromley Libraries, London Borough of Bromley.*

Printed in mid-1907 as a seven-page pamphlet, Willett's tract is a remarkable piece of work; without question, the London builder had spent a great of time developing his ideas and supporting arguments.[6] In sharp contrast to Franklin, who had urged people to rise with the sun, Willett argued that it was not possible to gain sunlight's benefits by appealing to early rising: "The exceptional exercise of this virtue usually calls forth more sarcasm than admiration or imitation." Moreover, said Willett, "Leisure must follow, not precede, work, and compulsory early business hours are quite unattainable." Having sunlight in the evening hours was the prerequisite.

Implementing his proposal seemed deceptively simple. At 2:00 A.M. on each of the first four Sundays in April, clocks' hands throughout the kingdom would be advanced twenty minutes; then, at 2:00 A.M. on each of the four Sundays in September, the hands would be retarded twenty minutes. Via a table of London sunrises, Willett demonstrated that every year his specific proposal was in effect would add 210 hours of daylight, all of it coming after 6:00 P.M. (see Figure 10.2).

Although this patriotic Englishman did not mention it, gaining 210 hours more light after 6:00 P.M. by advancing clock hands can be accomplished anywhere. That same amount will accrue at the equator, in Paris, in Europe, and throughout the Northern Hemisphere—indeed, any place south of the Land of the Midnight Sun. Only advancing the clock less (or more) than Willett's eighty minutes and/or altering the observance period will change this result.[7]

Where did Willett find this "wasted daylight"? Obviously, on any particular day the hours and minutes between sunrise and sunset are fixed. Thus, advancing the clock's hands to gain more light in the evening necessarily leads to later sunrises *by the clock*. Nowhere in his pamphlet did Willett mention this compensating effect.[8] Only in passing did he define "the clear bright light of early mornings, during spring and summer months" as wasted daylight. Even more telling, he saw increased morning darkness as a virtue, maintaining that those who went to bed after midnight and slept with open windows would find later sunrises advantageous.

Throughout his pamphlet Willett emphasized the benefits of additional light in the spring and summer evenings, noting that the extra eighty minutes a day was "the equivalent of a whole holiday every week," providing opportunities for additional exercise and recreation, thereby improving one's "health and strength of body and mind."

Willett did not neglect the economic benefit of capitalizing on the "wasted daylight" for illumination, calculating a saving of a tenth of a

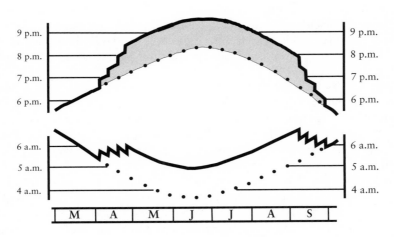

FIGURE 10.2

Willett proposed that public clocks be advanced four times in the spring and then retarded four times at the end of summer to return them to Greenwich time. Shown are the times of London sunrises and sunsets before changes in clock times (dotted curves), the later sunrises and sunsets after changing the clocks (solid curves), and the resulting increase in evening light (shaded area).

penny per person per hour—a total of £2,500,000 that would be saved every year by those living in the United Kingdom of Great Britain and Ireland. In a later edition of *The Waste of Daylight* he predicted that using less artificial light, and thereby reducing the amount of smoke in the air, would lead to "an appreciable benefit to eyesight generally, with correspondingly restricted need for the services of the oculist and the optician."[9]

For those Britons who might worry about the impact of his proposal on transportation, Willett had ready answers. Few trains are running at 2:00 A.M., so the only effect of advancing clocks by twenty minutes on each of four Sundays would be their arrival twenty minutes late on those specific days; otherwise their daily schedules would remain the same. Trains for the Continent, however, would require special timetables: one for April, one for May through August, and one for September. Willett dismissed the costs associated with producing these special timetables, arguing that the railway companies would more than make up the amounts in reduced illumination costs at the stations and also via an actual increase in rail traffic "as people are more ready to travel before than after sunset." (Subsequent testimony by various railway companies tended to bear out Willett's views regarding daylight saving time's minor impact on train schedules.)

Reminiscent of a long line of reformers, Willett was arguing that the benefits he articulated were great, while costs—scarcely estimated—were minimal. Further, his proposal for a periodic switch in time was easy to put into practice, despite the existence of *two* legal times for the United Kingdom.

As soon as Willett's pamphlet was in print, he began a lobbying campaign. From the very first he found support for his proposal from the editors of newspapers and periodicals, among them the *Spectator*, London's *Daily Telegraph*, the *Manchester Guardian*, and the *Liverpool Evening Express*. He also received encouragement from numerous well-placed individuals: clerics, educators, lawyers, scientists and engineers, and fellow members of the Royal Astronomical Society.[10] The idea of gaining additional sunlight during the spring and summer months had struck a responsive chord.[11]

Willett's venue was of course Parliament. On 4 February 1908 legislation embodying his ideas was introduced in the House of Commons via a private members' bill. By mid-March twenty members of the House of Lords and 175 members of the House of Commons had promised to support the legislation.[12] However, when the bill itself came before the House of Commons, it was greeted with laughter and jeers. The *Times* later claimed that its passage to a second reading was the result of a parliamentary accident.[13] In any event, the bill was referred to a nine-member Select Committee, which held its first evidence-gathering session in early May.

Willett was the first witness and summarized the letters and statements of support he had collected. However, some opposition to his proposal had appeared, and he was questioned about the extent of this resistance to advancing the time shown on clocks.[14] He replied that he had asked via the newspapers that objections be sent to him, indicated he had received none, and submitted a list of press articles that were both in favor of and opposed to his proposal.[15]

Over the next weeks the Select Committee met thirteen times, taking evidence from forty-three witnesses.[16] Most of the testimony was favorable, so much so that the committee chairman even asked Willett to advertise for opponents to the proposed measure.[17] Considering that the bill had been greeted with such derision in the House of Commons, the level of favorable support it was garnering undoubtedly surprised the government. Clearly, Willett had planned and executed his campaign extremely well.

What little opposition surfaced came from sections of the scientific community. The editors of *Nature* complained that the proposed "seasonal meddling of the clocks" was unjustified "from a scientific point of view." Instead of altering the clock, they supported a counterproposal for a change in people's habits—getting up an hour earlier in the summer—no doubt aware that would amount to continuing the status quo. They cited opposition to the bill from the astronomer royal, Sir William Christie, and from the astronomer Sir David Gill, without mentioning that the equally distinguished Oxford astronomers H. H. Turner and Arthur Rambaut as well as Sir Robert Ball of Cambridge had all spoken in favor of the measure; so had the world-famous Nobel Prize–winning chemist, Sir William Ramsay. The editors quoted the director of the Meteorological Office, who warned of considerable ambiguities in weather observations—observations that would become part of the kingdom's climatological records—that would result if Parliament shifted from Greenwich Mean Time to a varying set of times and then back again.

The editors of *Nature* went on to suggest that the proposal effectively meant adopting a dual-time system—which of course already existed with the United Kingdom's use of Greenwich and Dublin times. They used the false analogy that shifting the hours of daylight was the same as cutting an inch from the imperial standard of length to impress upon their readers what they termed a loss in the "sanctity of standards." And they criticized Willett as a member of the leisure class, contrasting the image of him breakfasting at nine with "the millions who begin work at six" for whom the shift in the hours of daylight would convey no benefits and, during many parts of the season, would force them to leave for work in the dark.[18]

Despite the editors' less-than-objective statements, a few of their criticisms reveal valid objections. Indeed, the problem associated with recording weather observations consistently was one nearly impossible to resolve. But these spokesmen for science were certainly biased. They did nothing to alter a misconception that arose early on and persisted for years: that "the hour of everything will be altered, and that everybody will begin everything an hour earlier."[19]

Also valid were certain of the objections raised during the Select Committee's sessions. Of the several railway directors who spoke in favor of the measure, by far the greatest concern expressed had to do with multiple shifts in the clock. They much preferred two shifts only, an hour in the spring and an hour in the fall. The secretary of the post office testified that local mails would be unaffected by the proposed change, but mail to and from the Continent would be delayed one day. And the London Stock Exchange registered strong objections, because the shift in time during the summer meant the loss of the one-hour overlap with American financial markets.

While those anxious to sell multi-clock-dial networks powered by electricity were in favor of Willett's proposal, the spokesman for the kingdom's traditional clockmakers was quite opposed, not only to the eight 20-minute increments and decrements, but also to the whole scheme of changing clocks back and forth. He countered by proposing that public time in Great Britain and Ireland be advanced one hour on a permanent basis.[20] Nevertheless, even those few opposing witnesses agreed that the opportunity to have more light in the evening was important for the public's general health.

The Select Committee submitted its report at the end of June. It concluded that changing clocks during April and September to provide more light in the evenings "is desirable and would benefit the community if it can be generally attained," noting that the weight of evidence supported that view. It discarded the multiple shifts given in the draft bill, and recommended that clocks be advanced one hour, at 2:00 A.M. on the third Sunday in April, and then retarded one hour, also at 2:00 A.M., on the third Sunday in September.

Then, in what can only be termed a blunder, the Select Committee, having dropped "Daylight Saving" in the bill's definitions as a response to comments that no daylight at all was saved by shifting hours, replaced it with "Local Time (Great Britain and Ireland)." Great was the consternation of scientists who pointed out that "local time" had a specific, defined meaning. No matter. Despite the strongly favorable report, the government was not inclined to take any action, with the prime minister, Herbert Asquith, refusing to grant time for debate.[21]

Up to now the *Times* had been content with keeping its readers apprised of the bill's progress and summarizing some of the testimony before the Select Committee. However, as soon as the committee's report was in print, the newspaper printed a lengthy editorial indicating its fundamental skepticism regarding the proposal. Since the bill did not make the switch in time mandatory, and believing that as a result many segments of society would continue to be guided by the sun rather than the clock, the editors wrote that "one part of the community will follow the real time, another part the fictitious time, and . . . worst of all, neither section will be able to follow either time with absolute consistency," leading to "intolerable confusion and dislocation." Then, after asserting that "the verdict of science is absolutely against" the bill and concluding that "[i]t is impossible to find sufficient justification for tampering with our [Greenwich time] standards," the *Times* editors let loose a flood of letters to the editor. These communications added little to the subject, for many were simply sparring bouts between proponents and opponents of the advanced time proposal.[22]

In 1909 daylight saving legislation was again introduced in the House of Commons, also as a private bill.[23] Now called the "Summer Season Time Bill," it incorporated the modification recommended in the Select Committee's 1908 report: to advance and then retard clocks by exactly one hour. This alteration reduced the number of additional hours of daylight after 6:00 P.M. from 210 to 154, a change opposed by Willett almost until the end.[24] During the bill's introduction, it was made clear that those who would benefit most lived and worked in cities and towns, the population ratio between them and rural dwellers being three to one.

Just prior to the introduction of the legislation, Willett sent letters to every member of the U.S. Congress urging them to adopt similar legislation. By doing so he hoped to continue the one-hour overlap between financial markets in London and New York—an action designed to counter the objection raised the previous year by the London Stock Exchange.[25]

When the bill came before the House of Commons for a decision on its second reading, a lengthy debate ensued. As before, several members of Parliament resorted to ridicule, which was countered by the president of the Board of Trade, Winston Churchill, who argued that the bill could not simply be laughed out of court as an absurdity. Reporting on the debate, the *Times* added a significant comment by noting that those who argued that moving the hands of the clock was the alteration of a chronological truth were ignoring the fact that "Greenwich time is itself nothing but a convenient 'make-believe' in every part of the country except [at] the precise longitude of Greenwich." Further, it wrote:

> If we have adopted one such convention for the sake of general convenience, there is no reason why we should not adopt another, *if we are convinced that it will be to our advantage*. (emphasis added)[26]

And although a strong majority (130 to 94) voted in favor of the bill's second reading, the *Times* also pointed out that there had not yet been a sufficiently broad discussion of the proposal by all segments of the kingdom, especially segments of the agricultural community. The newspaper applauded the decision to again refer the bill to a Select Committee.

The makeup of the second Select Committee was distinctly different from that of the previous one, with several of its sixteen members already publicly opposed to the proposed legislation; one sees in this a more carefully considered selection process by the Asquith government. So, with the favorable report of the first Select Committee already before them, the Select Committee deliberately—and quite correctly—focused on opposing views.[27]

The three key groups were agricultural interests—farming and dairy—newspapers, and public workers' unions. The bulk of the testimony came from the associated organizations, whose members had been polled. Countering testimony came from individuals—the fruit growers in the south of England and large milk producers that supplied London residents, the editor of a large metropolitan daily, and other workers' societies. Considerable evidence was presented, much of it conflicting.

Since the sun would be rising later by the clock under the proposed legislation, the effect of the morning dew on crops was considered. Fruit growers picked even when the dew was heavy and so saw no disadvantage to advancing the clocks, with the later evening light giving consumers an additional hour to purchase their daily shipments. Wheat farmers argued against the later clock sunrises, for they could not reap and stack if their crops were still damp; workers would stand idle in the mornings and demand extra pay for working beyond normal quitting times.

The kingdom's milk trade was a mixture of large urban suppliers and local ones marketing to smaller towns. The former did not care much about morning sunlight, for their breakfast milk was shipped in the night before. On the other hand, suppliers in smaller towns, who brought in the more-desired warm milk collected that morning, argued that their horse-cart drivers would have to rise even earlier than before if sunrise by the clock was later. Curiously, the cows' altered milking times was not seen as an issue.

Such technical nuances, many of which clearly exaggerated the damage or benefits that a change in the hours of light would bring, meant that the members of the Select Committee faced multiple quandaries. The committee tended to resolve these by accepting the positions of organizations representing multiple members while discounting the statements of individuals in the same occupations.

The Select Committee's evidence-gathering phase had started in April and was complete by early July. Twenty-four witnesses had testified, among them Willett, who spent most of his time contesting statements from the proposed legislation's opponents. Members met to consider their report to Parliament. The chairman drafted a favorable statement; however, one committee member wrote an unfavorable one. When the vote to accept the chairman's report came, an amendment calling for its replacement by the negative report was laid on the table. It passed six to five, and so the chairman's draft was rejected.

A flurry of action ensued. Days passed. Finally, a compromise was reached, and in late August a report, which had reportedly received nearly unanimous consent, was issued. It stated the committee's support for the principle of gaining more light during spring and summer evenings, expressed a wish that this be accomplished through voluntary means, and recommended "that the Bill be not further proceeded with." Insofar as Parliament was concerned, the issue had now been dealt with.

Despite this second defeat in the House of Commons, Willett forged on. In 1910 and especially in 1911 he arranged for demonstrations of support at London's Guildhall. At the first one the major speaker was astronomer Sir Robert Ball, a well-regarded, effective lecturer. At the second Guildhall meeting the featured speaker was Winston Churchill, now home secretary, who spoke to an enthusiastic capacity audience (see Figure 10.3).

Churchill reminded his listeners that four years had passed since the idea of daylight saving time had been presented to Parliament, and in the ensuing years more and more private organizations and citizens had voiced their approval of the measure. Now, after mature consideration, 170 city corporations and town councils, 46 chambers of commerce, 265 members of Parliament, 53 trade unions, and over 200 associations and clubs supported the principle of daylight saving time.

Answering those critics questioning the need for an act of Parliament to alter the kingdom's clocks in the summer, Churchill insisted that it was "impossible for individuals to alter their hours apart from the general movement of society without difficulty and inconvenience." He continued by noting that despite the inconvenience associated with

FIGURE 10.3

Winston Churchill, ca. 1910. Throughout the debates in Parliament, he was a staunch advocate of William Willett's daylight-saving-time scheme. Library of Congress, LC-B2-DIG-ggbain-04739.

individual action, many establishments throughout Great Britain were actually doing so, and the change in summer hours "was popular among all classes of workers wherever it had been adopted." Without any doubt, he asserted, a daylight saving proposal would eventually be adopted throughout the kingdom and then in other countries around the world.

As expected, Churchill's resolution in favor of a daylight saving time bill was seconded and carried unanimously. A resolution requesting the government "to give facilities for the Bill" also carried.[28]

Acknowledging the ever-increasing support that Willett's proposal was receiving and perhaps also concluding that if a Summer Time bill came to a vote in the House of Commons it would pass, the editors of the *Times* continued to argue strongly against altering Great Britain's standard time. Still in their minds was the view that the same result could be effected without recourse to legislation, for public institutions—government offices, the Bank of England, the stock exchanges, and banks throughout the country—could open an hour earlier during the summer months. Private businesses would soon follow, and then, necessarily, the public would rise an hour earlier. The editors insisted that advanced time legislation should be not passed until it could be shown to enjoy overwhelming public support and approval throughout the community— the equivalent of insisting on a supermajority in Parliament. Lastly, the editors opined, "If public opinion is ripe and eager for the change, no such Bill is necessary; if it is not, no such Bill ought to pass."[29]

The private bill introduced in Parliament in 1911 and subsequent ones introduced in 1912–1913 and 1914 went nowhere. On each occasion Willett prepared a new edition of *The Waste of Daylight*, each bulkier than the one before, and listing additional supporters.[30] Also in 1911 Willett pointed to reports by parliamentary committees in Great Britain, Canada, the state of Victoria in Australia, and New Zealand as evidence of worldwide interest; the following year he added German and French translations of *The Waste of Daylight* to the list.[31] In his last letter to the *Times*, printed in April 1914, this indefatigable advocate cited the support of 720 city corporations and towns and district councils as well as hundreds of societies throughout the United Kingdom.[32]

Ironically, gaining additional light in the evening hours came fourteen months after William Willett's death. World War I was well past the midpoint of its second year when, at 11:00 P.M. on Sunday, 30 April 1916, Germany and Austria-Hungary advanced all their clocks by one hour, shifting the official time of the Great Powers from Central European to Eastern European Time (see Figure 10.4). The ostensible reason for doing so was to reduce the quantities of petroleum oils, gas from coal,

FIGURE 10.4

Poster issued for the German-occupied areas of Polish-speaking Russia announcing the one-hour advance in Germany's legal time, beginning on 1 May 1916 and ending on 30 September 1916. Hoover Institution Archives, HIA GER-2809.

and coal-based electricity employed for lighting. Government officials claimed that the resulting economic savings would amount to millions of marks and millions of kronen.[33] Undoubtedly, the propaganda value of citizens changing their clocks in order to gain even minor savings as part of the war effort was also a factor.

By mid-May four countries had followed Germany and Austria's lead: occupied Belgium, which had been keeping Central European Time since late August of 1914; neutral Holland, whose standard time since 1909 had been twenty minutes in advance of Greenwich meridian time; and Denmark and Sweden, both of which had already adopted Greenwich meridian time, the former in 1894 and the latter in 1900. All advanced their clocks one hour until the end of September, a daylight saving period of five months.

The decisions made by the British and French governments in this war era are of special interest. In Great Britain, nearly a decade had passed since Willett had first introduced the subject; nevertheless, arguments pro and con were still being voiced. When, in February 1916, a member of Parliament again raised the subject, the government rejected taking any action. However, by early May its position had changed completely. The strong likelihood of strikes by the kingdom's coal miners and rapidly rising prices for coal had had an impact—as had the Central Powers' actions. In addition, the fact that the French government was also considering this issue changed the focus of the discussions.

Despite the looming coal crisis, the editors of *Nature* still opposed any time-change legislation. Again using a false analogy—"[i]t is usually understood that people cannot be made sober by Act of Parliament"— they insisted that the advanced time proposal was designed to make people "rise earlier by a legalised plan of national deception," reminding readers that they "had condemned this ridiculous measure whenever it has been brought forward." After the bill's introduction and the likelihood of its passage was clear, they grumbled that the fact that Germany had adopted the scheme was not a reason for Great Britain to adopt it, "but the reverse," and signaled their hope that France would reject the proposal.[34]

Debate in Parliament began on 8 May. The known opposition of farmers to advancing the clock was swiftly negated by the home secretary, who commented, "There are other people in the world besides farmers. Farmers work by sunlight, other people work by the clock. This Bill has been advocated as a war measure, and especially on the grounds of economy." That same day a vote was taken; the bill passed 170 to 2. Sent at once to the House of Lords, the legislation encountered no resistance of any consequence. The act received royal assent on 17 May.

Great Britain's first period of Summer Time began at 2:00 A.M. on Sunday, 21 May 1916, as it did in Ireland twenty-five minutes later. And when Summer Time ended on 1 October, clocks in Ireland were set back thirty-five minutes in order to conform to Great Britain's time (see Figure 10.5).[35] After thirty-six years, Dublin time was no more. For the very first time the entire United Kingdom was observing Standard Time, Greenwich meridian time.

It had taken no more than three months of debate and legislative pressure to effect altering the kingdom's clocks twice a year—a fundamental change in public timekeeping. The turmoil of the Great War and the possibility of savings on the home front had forced the legislation through.[36]

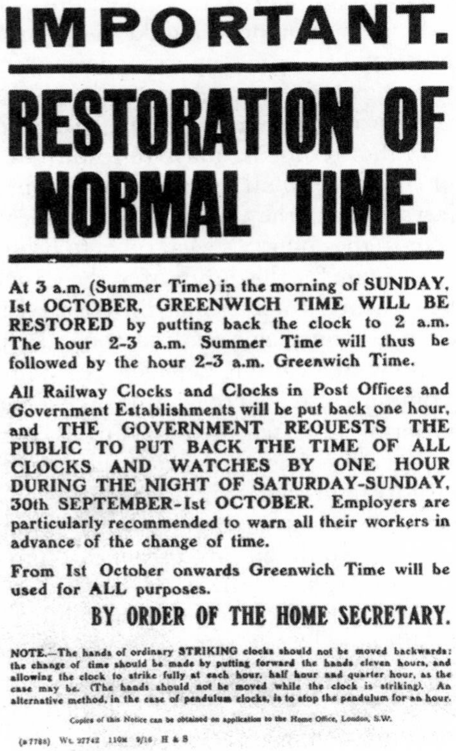

IMPORTANT.

RESTORATION OF NORMAL TIME.

At 3 a.m. (Summer Time) in the morning of SUNDAY, 1st OCTOBER, GREENWICH TIME WILL BE RESTORED by putting back the clock to 2 a.m. The hour 2-3 a.m. Summer Time will thus be followed by the hour 2-3 a.m. Greenwich Time.

All Railway Clocks and Clocks in Post Offices and Government Establishments will be put back one hour, and THE GOVERNMENT REQUESTS THE PUBLIC TO PUT BACK THE TIME OF ALL CLOCKS AND WATCHES BY ONE HOUR DURING THE NIGHT OF SATURDAY-SUNDAY, 30th SEPTEMBER-1st OCTOBER. Employers are particularly recommended to warn all their workers in advance of the change of time.

From 1st October onwards Greenwich Time will be used for ALL purposes.

BY ORDER OF THE HOME SECRETARY.

NOTE.—The hands of ordinary STRIKING clocks should not be moved backwards: the change of time should be made by putting forward the hands eleven hours, and allowing the clock to strike fully at each hour, half hour and quarter hour, as the case may be. (The hands should not be moved while the clock is striking). An alternative method, in the case of pendulum clocks, is to stop the pendulum for an hour.

Copies of this Notice can be obtained on application to the Home Office, London, S.W.

(a7705) Wt. 27742 110M 9/16 H & S

FIGURE 10.5

Home Office poster announcing the end of Summer Time in Great Britain and Ireland, 1 October 1916. Archives of the National Physical Laboratory, Teddington.

France took a different approach to the question of altering public time. Already the government had intervened with regard to public time-keeping: in 1891 and in 1911. On both occasions proponents and opponents argued passionately for their respective positions; obviously, many of those who lost in 1911 remained unhappy with the decision. Although government officials were aware of Willett's years of campaigning and the existence of persons in France favoring his proposal, the disinclination of Britain's Parliament to take any action from 1908 to 1912 certainly must have reinforced a view on the part of the republic's ministers not to intervene in timekeeping again.

Early in June 1914 a resolution supporting a report recommending that clocks be advanced—a report coauthored by Willett—was passed unanimously at the Sixth International Congress of Chambers of Commerce and Commercial and Industrial Associations, which was meeting in Paris.[37] But the assassination of Archduke Ferdinand at the end of the month and the outbreak of war in August made any legislative incursion into public timekeeping quite impossible.

However, by late 1915 the war in Europe had become one of attrition, a dreadful conflict requiring massive resources and the need for conservation and efficiency in all sectors. Hence, on 21 March 1916 a private bill was introduced in the Chamber of Deputies that called for an advance of one hour in the country's legal time, to begin at 11:00 P.M. on 9 April 1916 and continue for the duration of the struggle. According to its supporters, the measure would reduce the consumption of petroleum oils and coal used in illuminating cities and homes, objectives especially important to France, with its strong dependence on imported fuels.[38]

A ministerial committee created to examine the issues associated with the proposed legislation reported favorably, declaring that an advance in the country's legal time was not a scientific question, but simply a measure of social foresight that would lead to significant economies, thereby contributing to the national defense. The committee estimated that 130 million francs would be saved from the consequent reduction of artificial lighting in Paris, of which 80 million represented the government's own public lighting costs. For the country as a whole, the estimate was one billion francs. And, the report concluded, even if only a tenth of that amount was realized, a savings of 100 million francs was still impressive.

Additional supporting comments proffered during this period included a claim that the change could be effected easily, and that the advance of one hour would have little impact on the country's inhabitants. Further, the director of military railroads reported, the advance in time would cause few problems for the country's transport system.

Of course, objections were raised in the Chamber of Deputies, one member refuting the ease with which the change could be made by simply reminding his colleagues of the difficulties encountered after the passage of the 1911 legislation. He also questioned advancing the republic's legal time in the winter, noting the effect the later morning darkness could have on workers and children going to school.

While time-change discussions were taking place in the legislature, members of France's scientific community also considered the subject. One of the strongest objections came from Charles Lallemand, who sent a communication to the Académie des Sciences that was published in *Comptes rendus*.[39] A staunch advocate of the system of Greenwich-based time zones, Lallemand reviewed the situation in France. He began with the early-nineteenth-century change from apparent to local time, noted the late-nineteenth-century change to a national time—"legal time"—based on the meridian of Paris, and, of course, the four-year-old shift of legal time to conform to Greenwich time. He argued that in some areas of France the result of adding an hour to the current legal time would increase the difference between noon by the sun and noon by the clock to a degree unacceptable to the public. Lallemand's particular examples were weak, and his invocation of apparent time was perhaps a tactical blunder (because proponents of advancing clocks simply retorted that people no longer lived by the sun). But his view that too great a shift in the public's time would be resisted was a portent of the future.

Lallemand turned to the impact of the bill on various technical areas: tidal calculations and astronomical and meteorological observations. He predicted that a change to advanced time would cause great confusion in all areas—but particularly in navigation—when using tabulations that, of necessity, were printed months and years earlier than when they were used. Equally telling was his view that proponents of the change had greatly exaggerated the probable economic benefits, remarking that in many places, artificial lighting had already been reduced to a minimum. "It is truly the case in Paris," he wrote, "since the visit of the Zeppelins."[40]

Responding to Lallemand's communication, the Académie des Sciences directed that its astronomy section and its geography and navigation section meet together to study the issues. After doing so, the sections unanimously recommended that no change be made in the current mode of designating the hour, claiming—incorrectly—that the existing system was based on an international agreement that had been accepted after difficult negotiations conducted over a long period of time. Moreover, no time reform should take place until a preliminary agreement was reached among France's allies.

This technical position did not become official policy, however. On the same day that the sections' view was presented, a secret committee was established to examine the time advancement issue again. This behind-the-scenes activity eventually led the Académie to formally declare that "the [proposed] reform is of an economic and social nature . . . not dependent on scientific judgment."[41]

The question of advancing France's legal time also came before the Bureau des Longitudes. But after discussing the subject at its regular meeting, the bureau unanimously adopted a position drafted by Lallemand, concluding that the technical difficulties associated with advancing the country's legal time via legislation would offer no offsetting benefits. The bureau stated that its opposition to the proposed change was based on these technical difficulties. Further, if the reasons given were insufficient and the legislators still felt it their duty to adopt the legislation, no change should be undertaken before concluding "a preliminary agreement with the [almost thirty] States, allied and neutral, adhering to the International Conference on Time."[42] The government agency most concerned with the basis for France's time had spoken authoritatively against the change.

Other sectors more in favor of advancing the country's legal time then weighed in. These included the Chamber of Commerce of Paris and the Academy of Sports.

At first neither Lallemand's technical objections nor the positive views of the narrowly focused advocacy groups raised much debate among the general public. But the French press was quick to take up the issue of advancing the republic's clocks. Indeed, Renaud reported that from the very beginning, numerous articles had appeared in the daily newspapers. Gradually, public opinion changed from widespread indifference to a sharp split like that in the scientific community. "Thanks to the campaign in the press, perhaps for the first time a large audience had become familiar with the concepts relating to the means of indicating time," the French navy's hydrologist observed.[43]

The bill, along with its favorable report, came before the Chamber of Deputies in mid-April. Not surprisingly, members were divided on the issue. Anxious to pass the legislation, the minister of public instruction, fine arts, and inventions of interest to the national defense defended it "with the highest authority attached to him and his functions." His arguments included the need to conserve coal, large quantities of which were being imported from Great Britain, and to make sacrifices for the national defense. But his remarks were to no avail; the Chamber voted 267 to 211 to send the bill back to committee for further review. Instead, it agreed on a substitute bill directing the government to take measures to regulate public and private consumption, particularly with regard to products and raw materials from abroad.

Sent to the Senate, this draft bill was examined by a special committee of nine members that requested statements from those able to provide clarification of the issues. The committee heard from the minister of public instruction; the minister of the navy, who was assisted by Renaud; Charles Lallemand, as a member of the Institute of France; the director of railroads in the Ministry of War, assisted by the chief engineer of the Compagnie du Nord; and Charles Nordmann, associate astronomer at the Paris Observatory and a longtime advocate of the reform. After hearing them the committee was still not convinced of the practical value of advancing the nation's legal hour and, impressed by the many inconveniences such a change would cause, decided not to support the bill.

At this point the minister of public instruction asked to be heard. He presented a compromise: that the *heure légale*, as defined by the law of 9 March 1911, be advanced one hour, starting at a time to be determined by decree and ending on 1 October 1916. Finally, the impasse was broken. The compromise was made part of a draft bill and immediately given a favorable report.

The Senate began final debate a few days later. Although opponents continued to voice skepticism regarding the likely economic benefits, the telling point was that France's allies, Great Britain and Italy, as well as a host of neutrals, had already adopted a similar measure and were observing advanced time at that very moment. So the bill passed the Senate and was sent to the Chamber of Deputies, where it passed without discussion. When Summer Time began in France at 11:00 P.M. on Wednesday, 14 June 1916, with the clocks advanced one hour, it was a war measure—less a permanent change than an experiment motivated by patriotism and propaganda.

It had taken the French government three months to adopt the legislation, the time it had taken Great Britain to adopt a nearly identical act. However, unlike Great Britain, France never had an opportunity to examine the issues, nor to learn the positions of special-interest groups. Moreover, the French legislature had been extremely respectful of the arguments put forth by its scientific organizations, those closely linked to the government. More than likely, if the Italian and British governments had not already advanced their own clocks and thus negated the technical objections, the change in France would have taken much longer to effect. Small wonder the government minister compromised.

Like many of their fellow legislators in Britain's Parliament, many French deputies and senators remained skeptical of the economic and other benefits that proponents put forth as reasons for adopting advanced time. Before May 1916, of course, there were absolutely no data in hand, only cost-benefit estimates and assertions.

Soon after advanced time ended in France, *Le Génie Civil* began publishing accounts of savings in electricity in various parts of the country, attributing the reductions to increased natural light in the evening—and to the mandated early closing of cafés and other civilian businesses. Thus the journal's editors reaffirmed their support for *l'heure été*.[44]

As its own Summer Time was ending, Britain's home secretary appointed a committee to investigate the effects of increased light in the evenings. Undoubtedly established in anticipation of a government decision on Summer Time for 1917, the committee faced the dual problem of examining only slightly more than four months' experience, with a deadline for action looming. Consequently, its report, dated 22 February, cannot be considered authoritative in the sense of providing well-documented conclusions. Indeed, much of its content was based on questionnaire surveys of affected organizations and groups.[45] All the same, the quality of this report, which offered an objective view of the issues and freely acknowledged the uncertainties associated with its judgments, was excellent. It remains to this day a model to emulate.

Three key areas investigated were quantifying the saving from reduced consumption of coal and other fuels for lighting purposes, the effects on agriculture throughout the United Kingdom, and the consequences of Summer Time in Ireland.

As noted earlier, the impetus for Summer Time legislation had been anticipated savings in various energy sources during the spring and summer evenings. Consequently, the Summer Time Committee focused on determining how much was saved in the various energy sectors. Other factors—for example, blackouts in the cities, mandated earlier closings of public houses and many shops, and higher charges for electricity—also contributed to lower consumption during the observance period. This made it well-nigh impossible to reliably estimate the savings attributable to Summer Time. Indeed, in every case examined—fuel oil for illumination, coal and gas for lighting—the committee could say little more than the increased light in the evening "must have resulted in a considerable net saving."[46] Having completed this important if futile effort, the Summer Time Committee moved on to other issues.

As already anticipated, those engaged in agricultural activities in Great Britain turned out to be less enthusiastic than town dwellers about Summer Time. Questionnaire results showed that in many agricultural communities Summer Time was ignored, with farmers simply refusing to alter their clocks. This of course led to many complaints, for children in these communities had to go to school by Summer Time, even though their mealtimes did not coincide with those of their families. In addition, heavy dews on the fields caused a shift in the daily harvest time, leaving workers idle during the early hours waiting for the crops to dry out. The report further noted that many farm workers refused to work an hour later in the evening, or else demanded an extra hour's pay. Brushing these complaints aside, the report of the Summer Time Committee concluded:

In spite of such difficulties as have been recorded, a majority of farmers and War Agricultural Committees are in favour of the renewal of the Act, and the majority of even those [opposed] . . . consider that it should be renewed, as they recognize its great benefit to the community at large.[47]

They added that if altering Great Britain's clocks continued, the kingdom's farmers preferred a five-month observance period, one running from the middle of April to the middle of September.

The committee received conflicting reports from groups in Ireland. Many Irish town dwellers favored Summer Time; those engaged in agricultural pursuits opposed it. However, in Ireland these latter were more strongly against the time shift than were the agriculturalists in Great Britain. Consequently, in the committee's view, continuing Summer Time in Ireland was a more problematic action, especially in view of the just-inaugurated twenty-five-minute advance in Ireland's civil time, which produced later sunrises (and sunsets) throughout the entire year. Accordingly, the report recommended that future advances in time in the United Kingdom be carefully restricted so that summer sunrises in Ireland would not be too late by the clock. Specifically, it was proposed that Summer Time's observance period end no later than the third Sunday in September and begin no earlier than the second Sunday in April, again a five-month observance period.[48]

Despite its inability to quantify the effects of advancing the clocks, the Summer Time Committee wrote that it could "unhesitatingly say that the vast preponderance of opinion throughout Great Britain is enthusiastically in favour of Summer Time and of its renewal, not only as a war measure, but as a permanent institution." Concluding that sufficient advantages had been gained by advancing the clocks during the summer months of 1916, the committee formally recommended that the United Kingdom continue to observe Summer Time even after the end of World War I.[49]

Yet, a significant fraction of the population did not feel this way, so the history of daylight saving time subsequent to its adoption as a wartime measure reflects the tension between the two opposing views. In the case of the United Kingdom, Summer Time continued throughout the

1920s and 1930s, but with adjustments, some unsuccessful, to achieve uniformity in time with France and Belgium. The United Kingdom also experimented with different observance periods in an attempt to satisfy both urban and agricultural interests simultaneously.[50]

In a delightful and ironic ending to the early history of advanced time in Great Britain, William Willett was honored in 1927 by a group of

FIGURE 10.6

Tablet erected in 1927 in Petts Wood, near St. Mary Cray in Greater London, to honor William Willett. The numerals on the sundial have been advanced so that it always shows the hours of Summer Time. October 2001, by author.

Chislehurst residents anxious to prevent further encroachment on their suburban environment by developers. A large section of Petts Wood, where Willett used to ride, was purchased and deeded to the National Trust and renamed Willett Wood, and a tall granite tablet was erected as a memorial to this indefatigable proselytizer and propagandist (see Figure 10.6). A sundial was cut into the south-facing side of the memorial stone, with its hours advanced to show Summer Time; below it the inscription (in Latin) reads, "I will tell only the summer hours." On the north face were incised the words, "This wood was purchased by public subscription as a tribute to the memory of William Willett, the untiring advocate of 'Summer Time.'" The monument still stands.[51]

CHAPTER 11

Changing Time, Gaining Daylight

[T]he proposal so to arrange our standard time as to bring more
daylight into the end of the day is a suggestion . . . well worthy [of]
any expense or effort to effect its accomplishment.
—Chicago Association of Commerce, 11 December 1914

The experience of the United States between 1883 and 1926 illuminates
the fundamental issue associated with advanced time: the irresolvable
conflict between the desire for more light in the evenings throughout the
year and the resulting later morning sunrises. Although the issue was not
seen as central during the era's first decades, its recognition during the
first period of national daylight saving time transformed all subsequent
debates regarding changes in public time.

When in 1883 North American railroads adopted Standard Railway
Time, many municipalities quickly accepted the Greenwich-based system. The spread of one time for all civil purposes transformed Standard
Railway Time from a system encompassing narrow bands of railway
track to one throughout which large regions of the country had the
same time—now called time zones instead of time belts, the original
designation.

The places where the time shifted by one hour, which defined the
boundaries between zones, were those proposed by William Allen, who
selected them on the basis of his view of the railway companies' convenience. As a result, the area encompassed by what became the Central
Standard Time zone was huge, more than three times larger in area than
the more populated Eastern zone. (The former was 111 min in breadth—
a time span of nearly two hours—while the latter was 70 min wide.)

At first, those municipalities far from a zone's time-defining meridian refused to accept the new system, for the change from local time
to zone time had a significant impact on their clock times of daybreak

and sundown. Resistance was especially strong in cities situated in the eastern regions of the zones, among them Cincinnati (22 min from the time-defining meridian), Detroit (28 min), Savannah (36 min), and Bangor, Maine (25 min). Their complaints were identical: shifting to the new time would cause the clock times of their sunsets to be earlier.[1]

Over the next few years, several state governments adopted their respective zone's time as *legal* time, leaving municipalities free to adopt whatever time was convenient for them. When, in 1885, the Michigan legislature adopted Central Standard Time, Detroiters living at the zone boundary continued with local time—which meant that two times were still in use, but the local railroad time and state time were now identical. In late November 1900 the city council of Detroit decided to adopt Central time for its offices, to conform with the state ones. Immediately the question of what time would be used by the city's businesses, churches, theaters, and factories came to the fore. Should they continue keeping local time but open and close a half-hour later; or shift to Central time and keep the same hours? And what time should citizens keep in their homes?

Undoubtedly the sudden shift in sunset times was a significant factor in the controversy: On 20 November, the day the ordinance passed, the sun set at 4:35 P.M. The next day the sun set twenty-eight minutes earlier, at 4:07 P.M.—a most noticeable change. Protests began, time confusion reigned, and local newspapers had a field day.

Before the end of the month, the city council reversed itself and switched back to local time over the protests of those aldermen in favor of continuing with Central Standard Time. Moreover, it was pointed out by one of the local newspapers that "standard time is the legal time of the city and county and will continue to be until the [Michigan] legislature changes the standard," adding that many commercial and legal transactions had to conform to the official time and that the city council had no authority to pass an ordinance negating the use of that time anywhere within its jurisdiction. Confusion continued.

Proponents of the change persisted. After a popular vote in 1905, the city of Detroit adopted Central Standard Time.[2] But the issue of the clock times of sunset refused to go away. The Detroit More Daylight Club came into being, its goal an ordinance to advance the city's current time by one hour on a *permanent* basis.[3] Members lobbied the local government and newspapers. Via ordinance in 1915, the city government adopted Eastern Standard Time, an action subsequently approved by Detroit's voters.[4] Of course, the state offices continued with Central Standard Time, as did the railroads running to and from Detroit. For Detroiters, the net result was sunsets thirty-two minutes later than sunsets had been under local time (and, of course, one hour later than sunsets had been under Central Standard Time).

In addition to local agitation, attempts were made to turn the issue of earlier sunsets by the clock into a national one. In May 1909 a group of Cincinnati delegates representing the National Daylight Association met with their fellow Cincinnatian and newly elected president of the United States, William Howard Taft. The city had adopted Central Standard Time in February 1890, which, as a result of its being located east of the time-defining meridian, made its sunsets twenty-two minutes earlier by the clock.[5] Now, in 1909, the delegates were asking for the president's support in advancing time throughout the country by *two* hours during the summer months.

President Taft agreed to consider their appeal and met with his cabinet the next day. Already the superintendent of the U.S. Naval Observatory had been advocating a shift of one hour earlier in the Department of the Navy's official working hours during the months of June through September. However, cabinet members concluded that any shift of the nation's time would have a negative impact on mail schedules and national banking hours, so the executive decided against both proposals.

The National Daylight Association immediately turned to Congress, and within a week daylight saving bills were introduced, "by request," in both the House and Senate. The bills were modeled on William Willett's initial proposal of advancing clocks in four 20-minute steps during April and retarding them in four steps during September. After referring the bills to their respective committees, Congress took no further action— rather a foregone conclusion, considering the negative view expressed by the president's cabinet.[6]

Undaunted by this turn of events, the National Daylight Association switched its focus again to the city of Cincinnati. Late in June (1909) its city council passed an ordinance to advance city time one hour beginning 1 May 1910 and ending 1 October.[7] Thus Cincinnati became the first municipality in the United States, if not the world, to enact daylight saving time legislation.[8]

But it was not to be, even though advancing clocks during the summer months would have given Cincinnatians thirty-eight minutes of additional light during the evening hours than *before* the 1890 change to Central Standard Time. When the first of May arrived, the local newspapers carried not a single word about a shift in time. The ordinance had been repealed two weeks earlier.[9]

Local agitation—either for a permanent change or for summertime alterations to mitigate early sunsets—surely would have continued but for World War I, which changed the situation. National efforts now became the vehicle, the advent of daylight saving time in Europe the catalyst, and New York City the leading proponent for change.

For years many in the city's financial community had been advocating earlier business hours during the summer, and in May 1916 the same arguments for change were put forward once more.[10] By the end of the month, with Great Britain's adoption of Summer Time now in effect, Manhattan's borough president, Marcus M. Marks, established a working group, the New York Daylight Saving Committee, composed of businessmen and representatives of labor organizations. Its goal soon became one of establishing a nationwide movement to turn clocks forward one hour during the summer months.[11] Marks and his associates became leaders in the two-year lobbying effort that led to the adoption of Daylight Saving Time throughout the United States.[12]

During that same month came yet another indication of a desire to alter public timekeeping at the national level. Joint resolutions were introduced in both the Senate and the House authorizing the president to create a commission of government officials and other experts to investigate time standardization in the United States.[13] Clearly, this was necessary. Although for several decades the country's railroads and many states had been operating under a system of agreed-on times differing by exact hours, the system was not legally binding at the federal level. Unlike Great Britain, France, Germany, and Italy, among others, the United States had no national time.

The introduction of the joint resolutions in 1916 can be directly linked to a 1914 report prepared by the Chicago Association of Commerce. Its authors concluded that the country's standard time should be arranged "to bring more daylight into the end of the day."[14] Included was a discussion of the country's general adherence to the uniform system of standard time created by the railroads in 1883 and the statement that no change should alter that system. The report mentioned in passing the city of Cleveland's 1914 shift in time placing it in the Eastern zone, then went on to say that despite the one-hour difference between public time and the railroads' operating time, the resulting difficulties were merely inconveniences.[15] However, the report's thrust was that the permanent shift of one hour the association was advocating had to be done at the national level, "as local changes would tend to upset the now satisfactory and well-established standard time, upon which all railroad-time-tables and schedules are based"—a clear warning of opposition by the nation's railways to piecemeal shifts in time.[16] For over a year, copies of the text had been circulating in Washington.

On the afternoon following the joint resolution's introduction in the Senate, a short hearing took place before a subcommittee of the

Interstate Commerce Committee—an unusually rapid response to the introduction of any proposal. Only three senators were present at the hearing, and the sole witness was the Washington correspondent of the *Chicago Herald*.

The hearing's purpose was to consider the many facets of time standardization in the United States. Although the newsman noted the Chicago Association of Commerce's report, his testimony focused on recent European decisions to advance time by one hour. Calling their actions a "lesson which has been taught by the European war," he emphasized the distinction between neutral America and the warring countries. The latter had altered clock times to increase output, while Americans would gain an extra hour of light in the evening for recreation. His specific example was the baseball game that had been played the day before in Washington, which had been called on account of darkness—so different from home games played in Cleveland and Detroit, whose standard times had been advanced by one hour. He scarcely mentioned economic issues, save for the benefit to be gained from happier workers—ones in better health and thus stronger and working more efficiently—that would result from more light for recreation.

During his testimony the newspaperman included a historical summary of time in the United States starting in 1883. He noted Detroit's experience with its advanced time and submitted the text of the Summer Time bill that had just been adopted by Great Britain's Parliament. He proposed it as a model for Congress when it considered the legal issues that might arise from altering the nation's time.

After the newsman concluded, the chairman asked some minor questions, added some brief remarks, and adjourned the subcommittee, "subject to the call of the chairman."[17] No further action was taken by the Senate, nor by the House, so the joint resolutions died.

In the run-up to America's entry into World War I, Congress was undoubtedly facing issues more pressing than standardizing the nation's timekeeping. For all its inconsistencies, it had been operating rather well for more than a generation. Moreover, the legislators were being asked to address the matter primarily for recreational reasons. Already editorials against changing the nation's clocks had appeared, at least one of them voicing doubt that the European experience was at all germane to the United States.[18] Indeed, the spring of 1916 was probably too early to address public time at the federal level. Too wide a spectrum of views existed: make a permanent change, advance civil time only during the summer months, make no change at all. Couple this breadth of opinion with skepticism regarding the benefits that might accrue, Congress's usual reluctance to legislate in a new area, and no expression of interest at all from the executive. The sole recourse available to proponents of change was a strong public campaign designed to alter the environment.

Two chambers of commerce began advanced time efforts in 1916. During the summer Robert Garland (1862–1949), president of the Pittsburgh organization, took a unanimous vote favoring additional evening light to the Chamber of Commerce of the United States for its consideration. Responding, the national organization established the National Committee on Daylight Saving, with Garland as chairman and seven other members from chambers of commerce throughout the United States. In early December the national committee met in Pittsburgh and drafted a report strongly recommending legislation to advance the clock permanently by one hour—or from 1 April to 30 November if a permanent advance proved infeasible—with the intent of presenting this view at the annual meeting in Washington, D.C., scheduled for the end of January 1917.[19]

Also that summer the Boston Chamber of Commerce established a Special Committee on [a] Daylight Saving Plan. Chaired by A. Lincoln Filene (1865–1957), partner in the Boston firm William Filene's Sons, the seven-member committee began a study of the issues associated with advancing the clock. In August it distributed a thousand questionnaires in the New England area in order to gain the views of organizations, businessmen, professional men, scientists, and organized labor.[20] A month later it held a public meeting to publicize the results of the survey and to gauge public opinion.

Late in January the special committee completed its report and submitted it to the directors of the Boston Chamber of Commerce. The report contained a recommendation for a permanent one-hour advance in the nation's clocks (coupled with a fallback position to advance clocks only from 1 April to 30 November). Titled "An Hour of Light for an Hour of Night," the report was printed as a supplement in the chamber's 19 March *Current Affairs*. After a subsequent mail referendum among its members, advancing the clock permanently became the Boston Chamber of Commerce's official position.[21]

As already noted, Marcus Marks (1858–1934) became a seminal figure in the national campaign to adopt daylight saving time. A businessman who had retired to devote himself to civic affairs, Marks was already well known as a mediator and arbiter of industrial disputes. Using the offices of the New York Daylight Saving Committee, he enlisted the aid of numerous influential people. By mid-November plans for a convention to be held in New York City in January were announced.[22]

Early in January 1917 Representative William Borland of Missouri, who had been interacting with Marks, introduced the first of his three daylight saving time bills. All covered the two linked issues: establishing federal time zones for the United States and advancing the nation's clocks by one hour during the summer months. (A few days later Senator Gallinger of New Hampshire introduced an identical bill.) Coming near the end of the session, these bills had no chance of being examined by the cognizant committees, let alone enacted by Congress. They were simply placeholders, to be cited as the New York convention unfolded.[23]

Two weeks before the convention began, Marks and others spoke at a dinner for city publishers and editors at New York's Biltmore Hotel. Sponsored by the New York Daylight Saving Committee, which was seeking support for the House and Senate bills, this dinner led to newspaper coverage and publicity for the convention. A few days later, fifty members of the committee converged for a beefsteak dinner at which Marks and others spoke. This event was both a preparation for the convention and a reward for the rank and file's efforts. To ensure a continuing barrage of favorable publicity, a week before the convention Samuel Gompers, the American Federation of Labor's president, announced that the executive committee had passed a resolution announcing that it would give "greatest support" to the daylight saving movement.[24]

The convention was a great success. Held on 30 and 31 January at the Astor Hotel, sixteen hundred delegates from all over the country heard a host of advocates, among them Marks, the astronomer Harold Jacoby of Columbia University, Robert Garland of Pittsburgh, a member of Parliament who reported on Great Britain's 1916 experience with advanced time, and Representative Borland—who can perhaps be forgiven for declaring that the bills currently before Congress stood a good chance of passage. In addition, delegates were told of the support of the Federation of Labor, of the various daylight saving committees backing the convention, and of the activities of chambers of commerce throughout the United States.

At the working sessions the arguments presented in favor of advancing the clock one hour were the usual ones: reduced lighting costs, more daylight for recreation, better general health, greater protection of eyesight by not requiring laborers to work under artificial illumination, and a constant time relationship with Europe, whose nations would again be advancing their clocks.

Delegates spent much of their time considering when to set clocks forward and for how long. One delegate from Minnesota favored a four-month period, warning his fellow advocates that support from northern states would be lost if a permanent advance of one hour in time was adopted. A delegate from Detroit pointed out that the city had already shifted its time by one hour; he asserted that mothers there would rise in opposition if an additional shift of an hour came to pass, for the result-ing later summer sunsets would make it almost impossible to get their children to bed. The position apparently favored by most convention delegates was an hour's advance of five to five and a half months, similar in duration to the Summer Time periods adopted by both Great Britain and Germany during the previous summer.

During the convention's second day a letter from President Woodrow Wilson was read. Undoubtedly the conventioneers cheered on hearing the president's words: "I would have been glad to back up any movement which has the objects of the daylight saving movement." This statement was the first indication of support from the executive branch.

As the final order of business, convention delegates agreed to create a permanent organization, the National Daylight Saving Association, to lobby for national legislation (see Figure 11.1). Marcus Marks was elected president, with Robert Garland as first vice president.[25]

The next day (1 February) the Chamber of Commerce of the United States began its fifth annual meeting in Washington. Garland, chair-man of its Committee on Daylight Saving, presented the committee's report, which was accepted. Thus the official position of the Chamber

FIGURE 11.1

One of nearly a dozen and a half posters issued between 1917 and 1918 lobbying for the adoption of Daylight Saving Time. Library of Congress, LC-USZC4-8124.

of Commerce of the United States was to advance the nation's clocks one hour on a permanent basis. As an alternative, if a permanent change could not be achieved, the chamber put forward an eight-month advanced-time proposal, to begin on 1 April and end on 30 November.[26]

The last session of the 64th Congress ended early in March; on 2 April 1917 the 65th Congress began. Four days later the United States declared war on the Central Powers, a watershed for all domestic activities. Rapidly the executive began establishing control boards and regulatory bodies, and Congress passed a host of wartime legislative measures and sent them to the president for his signature. On 10 April Representative Borland introduced "A Bill to Save Daylight and to Provide Standard Time for the United States." Seven days later Senator William Calder of New York introduced S. 1854, an identical bill.[27]

The bill established five named zones for the United States and Alaska, linking their one-hour-difference defining meridians to Greenwich. Language in the bill directed the Interstate Commerce Commission to define the limits of each zone, "having regard for the convenience of commerce and the existing junction points and division points of common carriers," and giving the ICC authority to modify the zone boundaries from time to time. The section specifying how daylight saving time would be implemented followed the language of prior bills: At 2:00 A.M. on the last Sunday of April, the standard time in each zone would be advanced one hour, and at 2 A.M. on the last Sunday in September, civil time would revert back to each zone's standard time.

In May a subcommittee of the Senate Committee on Interstate Commerce held two days of hearings, amassing a sixty-six-page record of testimony and reports.[28] Not surprisingly, the testimony covered very familiar ground, for the vast majority of witnesses were longtime proselytizers for alterations in the public's time.[29] Two areas, however, were relatively new ones: the views of the country's railroads and the emerging national focus on food supplies.

Unhappy at the prospect of advancing the zones' times periodically, the railroad industry's technical expert on railroad timekeeping read the recent resolution of the Executive Board of the American Railway Association. It said in part, "[A]ny legislation excepting that which provides for a permanent change and which recognizes the present standard time zones shall be opposed on behalf of the railways." Conceding that Congress could pass legislation establishing or even altering the country's time zones, the industry witness pointed to the danger inherent in shifting railroad times twice a year, especially for operations on single-track lines. His aggregate figures were sobering: 1,698,818 trainmen's watches and company clocks had to be altered at the same moment twice a year to ensure correct time everywhere, and even at 2:00 A.M., a thousand passenger and five thousand freight trains were moving on the nation's rails.[30] However, advocates for advancing clocks had already cited railroad presidents by name who were on record as favoring changing times, and had done so recently. Thus the concerns of a technical body whose position had been formulated *before* America's declaration of war were simply entered into the record. How to actually implement such a time change simultaneously was, of course, not addressed in the proposed legislation.

Sensitive to the situation facing the United States at that moment, Marcus Marks had opened his testimony by noting:

The purpose that brings us here at this time, primarily, is President Wilson's message that the food supply needs increase. If we can have this daylight saving enacted . . . as far as New York City is concerned, there are a million people [who] come into New York, Manhattan, every day and go back to their homes every night; they will get there one hour earlier [*sic*] and they will have one more hour of daylight for gardening. Multiply that by the other cities in the United States, and the stimulation of gardening is tremendous.[31]

Marcus's other reasons for enacting the legislation, mentioned in passing, were savings in fuel costs and improvements in the public's general health. In neither his testimony nor his brief did he mention increased recreation opportunities resulting from an extra hour of evening light. The nation was at war, and he had seized on increasing food supplies, absolutely the best reason for daylight saving time, and, moreover, one never invoked before.[32]

Almost without exception, the other witnesses included increased food production in their testimony. Indeed, increasing foodstuffs via home gardening was now an official national goal, with advocacy efforts augmented by the National Emergency Food Garden Commission—later the National War Garden Commission—which had been organized in March by the extremely wealthy lumberman and conservationist Charles Lathrop Pack.[33] His successful propaganda efforts in favor of home gardening and canning and drying vegetables are the stuff of legend.[34] Pack had linked war gardens to the need for more light during the evening hours, and in their lobbying posters the advocates of daylight saving time linked the war's rifle with the garden's hoe.[35]

The Senate's Interstate Commerce Committee moved swiftly, issuing its report on S. 1854 two weeks after hearings had adjourned. The report recommended passage of the legislation, citing increased food production, comfort and convenience for workers and the public generally, and the savings associated with reduced need for lighting and fuels. That the railroads' objections were "not without some foundation and force" was recognized. Mindful of the burden that would be placed on railroads once the legislation passed, and aware that enacting any legislation could not happen until the summer—which would lead to *two* changes in railroad times within two or three months caused by advancing clocks in the summertime and then retarding them in the fall—the committee included an amendment making the act effective on 1 January 1918.[36]

In late June the bill was debated in the Senate in a rather perfunctory manner, amended as indicated, read a third time, and then passed.[37] The next day a message from the Senate arrived in the House announcing that the upper body had passed S. 1854, "in which the concurrence of the House of Representatives was requested." As soon as he could, Borland asked for and received permission to read the bill that he had introduced in April (H.R. 2609). In his remarks he declared that his bill was not a fad, because other countries had already adopted daylight saving time, which he characterized as a necessary war measure. Asked about the bill's impact on the railroads, he replied that no schedules would have to be changed, and, moreover, that the time zones the industry had created thirty-four years earlier would simply be adopted—a reasonable assumption on his part. Borland inserted a great deal of the now-familiar material supporting daylight saving time into the *Congressional Record*, all the while knowing that no action would be taken on his particular bill until the Committee on Interstate and Foreign Commerce had examined it and issued a report.[38]

A week later the Senate bill (S. 1854) was referred to the House Committee on Interstate and Foreign Commerce. And there it sat alongside Borland's H. R. 2609 until 6 October, when the first session of the 65th Congress ended.

Undoubtedly the House had more pressing business than considering bills seeking to alter the nation's time, a change that most probably would not come into force until spring of the following year. Moreover, some representatives were obviously opposed to the bill's provisions, and keeping a bill bottled up in committee was and is a well-known delaying tactic. Marks and others had already generated significant public support for the bill—Marks claimed the support of 150 newspapers and numerous organizations in his early May brief to the Senate—but to no avail. So the advocacy organizations continued lobbying via a barrage of posters, articles in newspapers, and letters to Congress.[39]

The second session of the 65th Congress began in December. The following February the House Committee on Interstate and Foreign Commerce submitted *Daylight Saving*, its report on S. 1854. In it the committee reprinted the complete text of the Senate report and proposed amendments.[40] Without giving any explanation at all, it changed the daylight saving observance period to start on the last Sunday in March and end on the last Sunday in October—*seven* months of advanced time, rather than the nearly universal five-month period.

One supposes that in private some individual representing one of the lobbying groups espousing a permanent advance in the nation's time, but publicly willing to compromise on a period of eight months, had compromised even further.[41] Of course, we shall never know; however, it is extremely unlikely that such a technical change was made by committee members on their own volition.[42] As we shall see, this alteration to the bill's provisions fertilized the seeds of destruction that bloomed in 1919.

After clearing the usual procedural hurdles, the Senate bill, along with the House's proposed amendment, came before the full House on 15 March 1918. Discussion regarding the change in the observance period was not very enlightening. To the only question regarding the inaugural date for daylight saving time, the bill's floor manager noted, correctly, that England would begin its observance on 24 March, adding, incorrectly, that Germany would begin on 15 March (actually it was 16 April).

Much material in support of the bill, including an official letter from Herbert Hoover, the U.S. Food Administrator, was placed in the House record. After some semihumorous remarks by a skeptical House member, a vote on the amended bill's passage was taken: yeas, 253; nays, 40; "present," 6; not voting, 134.

The next day the Senate concurred with the House amendments. The bill was then examined and signed by officials of the two chambers and received President Wilson's signature on 19 March. And so at 2:00 A.M. on Sunday, 31 March 1918, the United States advanced its clocks one hour (see Figure 11.2). A new era in America's public time was being inaugurated; proponents of the change proclaimed victory.

Adjustments to the realities of the new time were made even before clocks advanced. In order to meet the requirements of the federal statute, the city of Detroit, which had advanced its clocks in 1915 in order to have Eastern Standard Time, passed an ordinance making Central Standard Time the city's official time until the last Sunday in October—thereby allowing it to continue with Eastern time.[43]

With no time available to define boundaries between the standard time zones, the ICC froze the current ones and directed the country's common carriers to observe the requirements of the law by advancing their operating times on 31 March. Later it reported that the carriers had been successful, having made the switch in time "apparently without any confusion or accident."

FIGURE II.2

*Private-sector lobbying poster announcing March 1918 passage of Daylight
Saving Time legislation. Library of Congress, LC-USCZ4-10663.*

Some weeks after the nation's clocks had been advanced, the ICC
initiated a preliminary investigation of the country's system of time. It
found itself "unable . . . to arrive at a proper basis for defining the limits
of the first four zones," giving its reasons:

The existing zones, so far as the term "zones" can be applied to areas interlaced
by railroad lines, are so irregular as to preclude an attempt to define them even
approximately. The [mid-]meridian lines have been ignored as boundaries. Railroads and localities in many instances employ different bases of time. In many
cases railroads in the same locality use different time standards.[44]

Forthwith the ICC inaugurated a comprehensive investigation and held
hearings around the country, with railroad representatives and state and

municipal authorities in attendance, in order to arrive at definable zone boundaries generally acceptable by those most affected.

At the hearings the ICC presented tentative boundaries, basing them on well-articulated and reasonable criteria. These included placing the resulting shifts in time as close to the mid-meridians (halfway between the time-defining ones) as feasible, keeping the zone boundaries away from heavily populated centers insofar as possible, making the zones as compact and symmetrical as practicable, assigning great weight to existing state statutes and municipal ordinances characterizing public time, and keeping the railroads' existing boundaries in mind. Sensitive to people's habits—that more of their ordinary activities took place in the afternoon than before midday—the commissioners located the boundaries somewhat west of the mid-meridians to secure "the greatest amount of daylight for the active hours."[45]

The net result of the 1918 ICC hearings and ensuing discussions and decisions was a set of federally mandated zone boundaries different in many places from the ones accepted by the railway companies thirty-five years before under Standard Railway Time (see Figure 11.3).[46] To avoid any confusion, the new boundaries were slated to take effect on

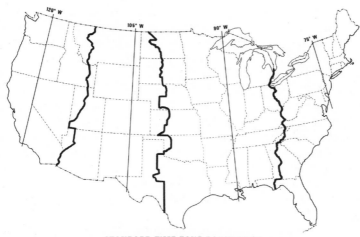

STANDARD TIME ZONE BOUNDARIES
AS ESTABLISHED OCTOBER 24, 1918 BY THE INTERSTATE COMMERCE COMMISSION
SOURCE: 51 I.C.C. 273.

FIGURE 11.3

U.S. time zone boundaries established by the ICC in 1918. From NBS "Review and Technical Evaluation," in Hearings on Daylight Savings *[sic]* Act of 1976, *Serial No. 94-109, 157.*

1 January 1919, a date deliberately chosen to be after 27 October 1918, the end of the country's first daylight saving period.[47]

While the ICC commissioners were considering the new boundaries, the Senate passed, with essentially no debate at all, a bill that would make the country's one-hour advance in time permanent. The request for this legislation had come from Bernard Baruch, chairman of the War Industries Board, whose justification for a permanent shift was that it would lead to "a very great saving of electric and other power."[48] Undoubtedly advocates for a permanent shift in zone time were trying to effect it as rapidly as possible, but their tactic failed. A month later, and less than three weeks after the ICC commissioners decided the country's zone boundaries, the Armistice was signed; World War I was over. Ten days after that the second session of the 65th Congress ended.

With the Great War now over, agitation began for the repeal of all war measures, including the repeal of the daylight saving section of the Standard Time Act of 1918. Moreover, the country's timekeeping environment itself was being buffeted by the federally mandated shifts in the time zone boundaries—alterations that in January would split the state of Ohio almost down the middle; partition the states of Virginia, Georgia, Florida, North Carolina, Oklahoma, and Texas; and make it impossible for Detroit to shift again to the Eastern zone. Just as significant, the views of one key group—the country's farmers—had not been considered during the many months of wartime lobbying for advances in civil time, for the advocacy efforts had focused on the advantages for urban and suburban workers and residents in the northern tier of states.

On 7 February 1919, three weeks before the end of the third and final session of the 65th Congress, the first of five House bills calling for repeal of the daylight saving provision of the act was introduced.[49] None stood any chance of becoming the subject of hearings and/or reports, let alone passage. However, their introduction signaled the unhappiness of a significant fraction of the country's population.

The 66th Congress of the United States convened in mid-May. During the preceding six weeks, citizens of the United States, many of whom in January experienced a one-hour shift in their civil times, now experienced an additional one-hour shift in their sunrises and sunsets, scheduled to continue for seven months. Obviously many more citizens were unhappy now, especially rural dwellers, whose occupations were more attuned to the rising and setting of the sun than to the clock. Stripped of the entertaining rhetoric generated during this period, the problem being addressed was later morning sunrises stemming from the two federal actions. Farm interests, and later many rank-and-file union members, including coal miners, complained to their elected representatives, and Congress responded.[50]

Early in the session a bill was introduced in the House to repeal the daylight saving time section of the 1918 act. Hearings were held in early June, and a favorable report was issued shortly thereafter. On 18 June the House debated the bill and passed it, referring it to the Senate the next day.[51]

Simultaneously with this unusually rapid legislative process, the House attached a rider to the Department of Agriculture annual appropriations bill, a required piece of legislation. Its language was designed to repeal the *entire* Standard Time Act—in terms of national time uniformity, a case of throwing out the baby with the bathwater. Majorities in both houses passed the appropriations bill and sent it to President Wilson for his approval. On 11 July he vetoed it, giving as his reason the section that would abolish daylight saving time in the United States. The House tried but could not muster the required two-thirds majority needed to override his veto. (The net result was the introduction of a second appropriations bill for agriculture with the offending clause removed, which passed both houses and was signed by the president.)

However, the battle was not over. In short order the Senate reported favorably on the House-approved bill, passing it on 1 August. Sent to the president for his approval, Wilson once again vetoed the repeal legislation. This time, however, both houses overrode his veto. Thus, at the end of October, when the nation's clocks were retarded one hour, national daylight saving time came to its final end—apparently.

But the campaigners for more daylight in the evening were not defeated. They took the battle to the state legislatures, and by the mid-1920s about a dozen states were advancing clocks for five months, May through September, with observance determined by each local jurisdiction. That the states could legally take such steps even though a federal statute regulating the country's civil time was upheld by the Supreme Court in 1926.[52] During the ensuing decades, northern states stretching from New England through Illinois and Wisconsin observed five months of daylight saving time—often, however, only in urban areas.

Local option in the observance of daylight saving time destroyed any hope for time uniformity in the United States. Not only were regions within the same zone keeping different times, but the observance period itself varied. Certainly for astrologers—and other groups interested in the exact time of an event—determining the time of a twentieth-century person's birth became difficult, a task assisted only by less-than-perfect compilations of changes in time.[53]

Alterations to the boundaries between the time zones provide a significant and fascinating coda. Among the earliest was the congressionally mandated 1921 removal of the Central–Mountain boundary from Texas and Oklahoma—undoubtedly to the pleasure of their respective state governments. Over time other shifts took place, albeit somewhat grudgingly, after hearings before the ICC. In this manner Ohio and most of Michigan were shifted to the Eastern Standard Time zone.[54] Most of the shifts were westward ones, continuing echoes of the late-nineteenth-century desire to increase the amount of light in the evening hours, for these states and regions were now west of the zone's time-defining meridian. (In today's parlance we would say that the westward shifts gave the various regions year-round daylight saving time, or YRDST.)

The decades between the 1920s and 1966 can be summed up as a struggle to regain uniform time for the United States, a goal that as of this writing (2006) has not yet been fully achieved. The road was a tortuous one, and anecdotes abound.[55]

Standard Time and Daylight Saving Time are inextricably linked. Some current issues associated with this fact are discussed in the epilogue.

Epilogue: The Present

☽

Uniformity is a constantly shifting, unattainable goal. Nowhere is that more evident than in the realm of public timekeeping.

To be sure, time uniformity is an absolute requirement for many endeavors. Pilots of commercial aircraft and the ground controllers who guide them all operate on Coordinated Universal Time (UTC), today's technologically advanced replacement for Greenwich Mean Time. For airline passengers, of course, that time has little meaning, for their plane's UTC-based arrival time is decoded by the cabin staff in terms of zone time at the landing field.

Space missions and military operations are also planned and executed in terms of a single time. Count von Moltke was absolutely correct when, in 1891, he urged adopting uniform time for imperial Germany's railways, with efficient troop mobilization as his example. Today, America's military not only reckons time in terms of UTC (called Zulu time), but employs a twenty-four-hour clock, designating hours and minutes from 0000 to 2359, while civilian government and the general public continue to embrace the old, familiar double-twelve notation. Yet, occasions arise when back-and-forth translation between the two is crucial.[1] In America at least, these two ways of specifying exact time will continue into the foreseeable future.

Changes in uniformity sometimes arise from altered political circumstances. After more than three-quarters of a century, the International Date Line, that seemingly immutable border between today and yesterday, shifted.

The Republic of Kiribati straddles both the equator and the date line and extends over a considerable area of the Pacific Ocean—across the equivalent of three time zones. An independent nation since 1979, Kiribati's territory includes the Gilbert Islands, most of the Phoenix Islands, and the Line Islands. Its population is slightly over 100,000, and of the

thirty-three low-lying islands that comprise the nation, twenty-one are
inhabited.

The anomaly of people living in one country but observing different
days ended on 1 January 1995. On that day, new time legislation became
effective, and since then all activities have had to be recorded in terms of
the Eastern or Asiatic dating (see Figure E.1). Those residing east of the
now-obsolete date line lost a day, with their 30 December 1994 being
followed immediately by 1 January 1995.

Kiribati's shift to Asiatic reckoning throughout its territory had an
unanticipated consequence. It put the country in the running to be the
first in the world to see the dawn of the new millennium. Other competi-
tors included Tonga, New Zealand, and Fiji.[2] Tonga uses the Asiatic dating, and since 1961 its official time has been thirteen hours in advance
of Greenwich time, even though it lies within the standard time zone

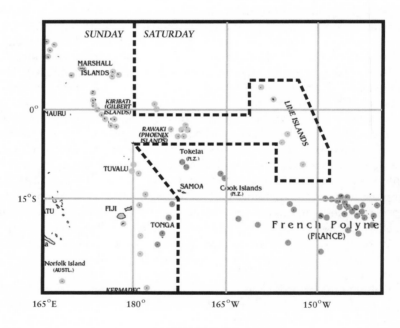

FIGURE E.1

*In 1995 the Republic of Kiribati, which straddles both the International
Date Line and the equator, declared that henceforth all its islands would use
Asiatic dating. As a result, a section of the boundary between the Asiatic and
the American dating shifted to the east, as shown in the figure. The base map
is from "Standard Time Zones of the World," 8031 105AI (R02183) 11-04,
published by the CIA.*

that is only twelve hours ahead. Late in the 1990s Tonga began observing daylight saving time (between October and March), a circumstance unusual for a country lying in the tropics. Thus, although it is located far to the west of Kiribati's easternmost islands, Tonga's new official time meant that the millennium's first moment would come to Tonga before Kiribati, by the clock.[3]

The Republic of Kiribati's formal change of its date reckoning is not shown on many maps. For example, the latest issue (2003) of the CIA's "Standard Time Zones of the World" continues the practice of showing the date line's sweep as it was before 1995. Yet ignoring the shift caused by Kiribati's decision effectively defeats the line's function, namely, to offer a convenient way of showing the boundary between the Eastern date and the Western one. This sole purpose was underscored by George Davidson more than a century ago. Further, appending "International" to "Date Line" is misleading, for it implies that the line cannot be redrawn until all—or at least some—maritime nations agree on its location. Clearly, this is not the case.

As shown in chapter 1, the misnamed International Date Line plays no role in ocean (or air) navigation, does not depict the limits of any nation, and has never been the subject of any international agreements. Unfortunately for map users who need to be aware of the precise boundary between the reckoning of civil days, the current nonuniformity in time zone maps and charts issued by government agencies and producers of commercial atlases will probably continue for many years.

By the eve of World War I, all the world's major hydrographic products had been linked to the Greenwich meridian (called the *méridien international* in France). However, land maps still continued to use a variety of initial longitudes. This was especially true for the large-scale ones published by national governments. As we have seen, such maps were always excluded from uniformity discussions held at international geographical congresses. This decision merely reflected practical reality: once a particular map series had been inaugurated, governments were most reluctant to alter it. When a new series was launched, more often than not the established framework of past map sheets was continued.

Some movement toward greater uniformity began after World War I. Italy, for example, inaugurated a 1:25,000 topographic series, with the meridian of Greenwich as its basis and the vertical sheet margins (neat lines) in terms of Greenwich longitudes. However, when Greece began its 1:25,000 series of topographic maps in the 1920s, the initial meridian was Athens.

Events after World War II led to major shifts in Europe's large-scale topographic mapping. Agreements among the NATO allies led to a high level of uniformity, including common projections, geodetic bases, and sheet parameters, and with Greenwich as the initial meridian. France's large-scale topographic map series, however, continued to be based on the Paris Observatory meridian, with a Greenwich-meridian scale outside the sheet margins.

Nonetheless, the latest printing of these French maps has two graticule systems for latitudes and longitudes: one printed in black in terms of France's system of grads and the Paris Observatory meridian, and one in light blue based on Global Positioning System values. Thus, the lack of uniformity so common throughout the nineteenth and much of the twentieth century is slowly vanishing.

The system of twenty-four 1-hour time zones circling the globe, all linked to the Greenwich meridian, is the culmination of over a century of country-by-country actions. Yet, from the start, exceptions to strict uniformity have persisted as a few nations have opted for a Greenwich-linked hour plus some fraction (usually one-half hour). Moreover, as we saw on the theoretical time zone map for Europe (see Figure 8.4), no one expected that a zone boundary would strictly follow a longitudinal line. Instead, the shift in time would usually occur at national borders.

Certain other changes have also reduced strict uniformity. After World War II, for example, France, Belgium, the Netherlands, and Spain advanced their official times by one hour, thus replacing Greenwich time with Central European Time.[4] It is most unlikely that these countries, all members of the European Union, will ever return to their "proper" zone time. Indeed, there is pressure within the United Kingdom to drop its long-held Greenwich time and adopt Central European Time.

Perhaps the only move toward greater uniformity in zone-based time will be in those places where civil time is not a multiple or half-multiple of an hour, but a fraction: Chatham Island (twelve and three-quarter hours in advance of Greenwich) and Nepal (five and three-quarter hours in advance). In any case, if one needs to be absolutely certain of the current time in some particular country, the surest way is to telephone a city hotel there.[5]

Almost one hundred years have passed since William Willett proposed altering clock displays to gain additional light during summer evenings. Welcomed by some and opposed by others, his proposal was adopted by many countries during World War I. During World War II, clocks in many countries were advanced one hour throughout the year, with some governments even adding an additional hour during the summer months. Once peace returned, opposition to changing public clocks twice a year resurfaced. However, sentiment for having more light in the evenings led the vast majority of countries in the Northern and Southern Hemispheres to adopt Daylight Saving Time (Summer Time).[6] Today, seventy-odd nations have adopted some form of advancing public clocks periodically.

In the United States, demands by the transportation and communications industries for consistency in civil time led to the passage of the Uniform Time Act of 1966. It cannot be emphasized too strongly how important this legislation was for establishing uniformity in public time-keeping. For the first time since World War II, all states and municipalities observed a mandated (six-month) DST observance period: from the last Sunday in April to the last Sunday in October.[7]

After the act took effect, a few states opted out, the most famous being Arizona and that portion of Indiana in the Eastern time zone. (Michigan, North Dakota, and Texas shifted their zone boundaries eastward to compensate for the combined effects of standard time and DST.) However, the goal of uniform time throughout the continental United States was essentially met. This situation changed only slightly in the ensuing years.[8] Indeed, the level of uniformity across the country actually increased.[9]

Although uniform time was achieved, the desire for longer periods of DST continued, and events led to changes in its observance. For example, as a direct result of the 1973 oil embargo crisis, a one-year period of DST took effect in 1974. This was followed by a somewhat extended period of DST in 1975. In 1976, the country reverted to its pre-embargo norm: six months of advanced time.

Proponents of additional DST spent the following years trying to pass legislation directly. After several failed attempts, they succeeded in having a rider attached to a certain-to-pass bill, and in 1986 the country's DST observance period changed from the last Sunday in April to the first, a three-week increase.[10] A decade later, the Energy Policy Act of 2005 extended DST observance an additional four weeks by having it begin on the second Sunday in March and end the first Sunday in November. This extension will take effect in 2007. In debates on all the bills introduced in Congress during these twenty years, the primary reason given for extending DST has been energy savings, in particular reducing electricity consumption because of more light in the evenings.

Without doubt, efforts to extend DST's observance period will continue. The northeastern states have been leaders in all past efforts. Short of switching to the Atlantic Time Zone (a move that was recently debated by Maine legislators), those northeasterners who want more light in the evenings can do nothing but lobby for longer and longer periods of DST.[11] Lobbying, of course, means emphasizing a national, not regional, goal to win passage of federal legislation. Also, without doubt, those living on the western sides of any time zone will, as they have in the past, oppose further extension of the DST period.

Astronomical, geographical, and demographic data were all included in a 1976 study by the National Bureau of Standards for the chairman of a subcommittee of the House of Representatives. This study gave Congress for the first time a systematic way to analyze the trade-offs associated with changes in Daylight Saving Time.[12]

Two of the summary charts and a table from this study, updated to reflect current information, are reproduced here. The first (Figure E.2) shows the envelope of sunrise and sunset times for the United States in 2006 under the current advanced-time statute. The second (Figure E.3)

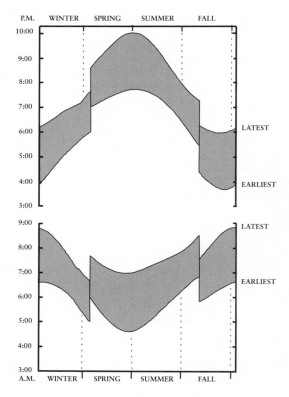

FIGURE E.2

Curves showing the range of sunrise and sunset times in the continental United States throughout the year. The shifted segments depict the observance of Daylight Saving Time, currently (2006) the first Sunday in April until the last Sunday in October.

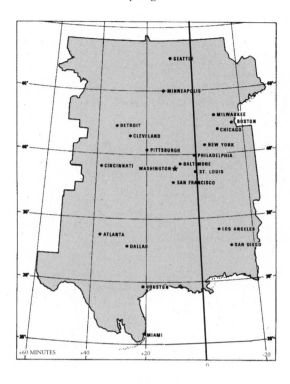

FIGURE E.3

The United States as a composite zone. The four time zones have been stacked along their time-defining meridian (heavy line) and the outermost segments of the set of boundaries drawn to show their geographical extent. The resulting zone ranges from thirty-two minutes on the east to sixty minutes on the west. The base map is from the 1976 NBS Report, with the western boundary adjusted to reflect zone changes in Indiana in January 2006.

is a composite of the four time zones illustrating that the spread of times about the time-defining meridian is not symmetric but extends from thirty-two minutes on the east to sixty minutes on the west. These charts help in comparing specific sunrise and sunset times. The population table (see Table E.1) quantifies the fact that significantly more people live to the west of the time-defining meridian than to the east.[13]

Questions about both energy savings and the safety of school-age children were raised during the debates that led to the adoption of year-round DST in 1974 and were examined in the 1976 NBS report just noted. The conclusions, based on the same data that the Department of Transportation collected and used in its 1975 report, were sobering. In the case

TABLE E.1

Distribution of population (in millions), continental U.S. time zones

| | Minutes from time-defining meridian | | | | | | |
	+60	+45	+30	+15	0	−15	−30	−45
Eastern	1.679	33.995	36.521	27.758	30.676	4.992	0.034
Central	1.267	8.611	23.784	18.521	27.139	2.187	—
Mountain	0.570	1.048	7.717	4.304	3.368	0.026	—
Pacific	—	—	.430	20.186	23.121	1.649	—
Total	3.52	43.65	68.45	70.77	84.30	8.85	0.03
Percentage	*1.3*	*15.6*	*24.4*	*25.3*	*30.2*	*3.2*	*0.0*

SOURCE: Population data from U.S. Census Bureau, 2000 census.

of electricity savings, the NBS report demonstrated that nationally no electricity savings could be detected. Regarding the added risks for fatalities among schoolchildren as a result of motor vehicle accidents on darkened mornings, the NBS analysts concluded there was a "statistically significant increase in school-age children fatalities in the morning" for the DST months of January and February 1974 compared with the same months the year before. (In its summary the NBS warned that there was no way to ascribe the increase to a "DST effect," considering the limited amount of data and the many factors that affect fatalities.)[14]

What is equally sobering and somewhat disconcerting is that in the succeeding thirty years, *no* authoritative federal study was ever undertaken to determine whether any DST-related energy savings actually exist. Nor has there been a federal study to determine whether the number of morning fatalities among school-age children is indeed related to extended darkness, especially around the dates of transition to and from DST.

In the case of the Energy Policy Act of 2005, general skepticism regarding energy savings may have led to the insertion of a subsection that directs the Department of Energy to study and "report to Congress on the impact of [the change in DST observance] on energy consumption in the United States." Clearly, this study will provide a useful and most welcome advance in our knowledge. However, obtaining clear results may prove elusive. The country's energy sources are diverse, and any energy savings, if they exist, are likely to be small and consequently difficult to quantify with certainty. For example, the demand for electricity, most of which is generated by coal and nuclear plants, is estimated to drop by 1–2 percent as a result of more light in the evenings. But the added period of DST will coincide with the heating season, so the increased consumption of natural gas and heating oil, a small portion of which is used to generate electricity, may overwhelm any energy savings from less use of electric lighting. Finally, increases in gasoline consumption have always been posited, based on the view that more driving is likely to occur when there is more evening light.

Concerning the safety of schoolchildren, in 2001 the DOT stated, "Based on the Bureau's [NBS] findings, the Department subsequently took the position that daylight saving time in January or February might possibly increase school age fatalities in the morning."[15] Since sunrise times in November and December are similar to those in January and February, DST could pose a problem during all four of these months. It must be noted that the Energy Policy Act of 2005 will extend DST observance in 2007 well into November.

Somewhat surprisingly perhaps, this 2005 amendment to the Uniform Time Act of 1966 may have little effect on uniformity within the United States. For example, states split by a zone boundary can petition the DOT to switch the affected regions (counties) currently in the more eastern zone to the western one, thus trading the late sunrises they have been experiencing for earlier ones. In this way they could continue observing DST while at the same time reducing their worries about the safety of children going to school in morning darkness.[16]

Yet, it must be remembered that uniformity in timekeeping extends to the entire globe. In the best of all possible worlds, all countries in the Northern Hemisphere's temperate zone would begin and end DST on the same day. Currently there is congruence between Canada and the United States (except in those regions that do not observe DST). Observance periods in the United Kingdom and the European Union are almost identical to those in Canada and the United States, with Europe starting DST on the last Sunday in March and the two Western Hemisphere nations on the first Sunday in April. All end DST on the last Sunday in October. While annoying, this one-week difference is relatively easy for the public to comprehend, and they can adjust their activities accordingly.

The four-week extension of DST slated to begin in the United States in March 2007 may have a detrimental impact on world time uniformity. Already concern has been expressed in Canada. If that country continues to follow the practice of its southern neighbor, as it has for decades, then it will fall further out of synchrony with Europe.

Moreover, it is very unlikely that European nations will modify their DST observance period to conform to the United States, or even be able to consider it by 2007. The likely result will be a period of time—years?—with the United States as "odd man out," leading to debates in Congress underscored by complaints from affected industries. With some luck, however, such debates may ultimately lead to even greater global time uniformity. For example, one compromise acceptable to many groups might be for the United States to begin DST on the same Sunday in March as the Europeans.

In the United States, a linked system of Standard Time and Daylight Saving Time has existed for four decades—and for up to nine decades in other places around the globe. Without doubt, this system will persist as the world's public time. Hopefully, its level of acceptance will continue to grow.

Notes

Introduction

1. Richards, *Mapping Time*, 24–25, 387–88.
2. This adjustment reduced, but did not eliminate, the difference between the Gregorian calendar and the length of the astronomical year; however, the former will not differ by one day until AD 3719.
3. Kelsey, "Gregorian Calendar," 239–51.
4. Richards, *Mapping Time*, 112–23, 239–52.
5. In his 1712 account of finding Alexander Selkirk, the model for Defoe's fictional character, Woodes Rogers writes that the marooned mariner diverted himself by marking on trees "the Time of his being left and Continuance there"; see the reprint of *A Cruising Voyage Round the World*, 94. The man of letters Richard Steele interviewed Selkirk soon after his 1711 return to England and wrote that he was once struck senseless but recovered his reckoning by observing how the moon had changed during the three days he lay unconscious; see Steele, *Englishman*, 108. This part of the story—of the isolated Selkirk regaining his daily reckoning via lunar observations—almost certainly is an embellishment.

Chapter 1

1. A useful summary of events is Mitchell, *Elcano*, 82–86.
2. Pigafetta, *Magellan's Voyage*, 2:185; Albo, "Log-Book," 235.
3. "Pigafetta's Account," 161. Undoubtedly Pigafetta was alluding to Gasparo Contarini's explanation as reported by Peter Martyr (see nn7 and 8 below).
4. I am deliberately ignoring one part of Oresme's thought experiment: having the travelers depart on the same day and moving around the globe at the same speed in order to arrive back on the same day. The philosopher includes this detail to show that the travelers' *count* of days will also differ. This additional constraint has, of course, no effect on the actual gain or loss of a day.
5. Lutz, "Fourteenth-Century Argument," 70.
6. Martyr's history of Magellan's circumnavigation was first printed in 1523 and was printed again in 1530 as the fifth of eight decades (see n8).

7. For an excellent analysis of Martyr's faulty reasoning and a summary of Contarini's explanation, see Selga, "Philippines," 4–6.

8. Martyr, *De Nouo Orbe*, fifth decade, chapter 7, 218. This 1612 work is cited as the first one in English giving all eight decades, the first three having been translated by Richard Eden in 1555.

9. Drake, *World*, 84; this account was written by Drake's nephew. Kelsey, *Sir Francis Drake*, 204, comments that the precise date of Drake's return is not clear, but that most historians accept 26 September 1580 (O.S.).

10. An important study of dates for the region is Selga, "Philippines," 4–15. Selga cites many of the era's Spanish sources.

11. Schurz, *Manila Galleon*.

12. Selga, "Philippines," 6–7, who embellishes somewhat the story of Fr. Sanchez. The precise year of the incident cannot be determined because the two feast days fall on the specific dates given here, and the original source—first published in 1590—does not give the day of the week; see n14 below.

13. Lamont, "International Date Line," 347. This work is cited frequently by Selga.

14. Acosta, *Natural and Moral History*, 172–74.

15. Carletti, *My Voyage*, 102–4.

16. Published in London in 1619 as *A Wonderfull Voiage Round About the World* and incorrectly ascribed to Schouten, the discussion, on p. 81, includes days of the week inconsistent with dates; e.g., "Munday the first of November" is actually Tuesday, 1 November 1616 (N.S.). In the Dutch and French editions of Jacob Le Maire's *Spieghel der avstralische navigatie*, printed in 1622, there are no days attached to the dates, indicating insertions by the publisher of the 1619 English edition. A brief quotation based on some edition of the English version is in Leigh-Browne, "International Date Line," 302.

17. Gemelli Careri, *Voyage*, 418–19; note that the header on the cited page is given as 1617 instead of 1616. Gemelli Careri's description of the horrors of the six-month and longer voyages from Manila to Acapulco on the Manila Galleon has been cited often. See, e.g., Schurz, *Manila Galleon*.

18. Dampier, *New Voyage*, 255–56. Note that Dampier's observation demonstrates the Spanish authorities' lack of influence in Muslim areas of the Philippine archipelago. Dampier also found that the Spanish at Guam were keeping the same Western reckoning as his; however, he did not know how it was being kept in the Spanish colonial regions of the Philippines, Manila in particular.

19. Rogers, *Cruising Voyage*, 285. Thomas, *True and Impartial Journal*, 179; Thomas was teacher of mathematics on the warship *Centurion* commanded by Anson. The change in dates made by James Cook are from the explorer's own accounts, as edited by J. C. Beaglehole.

20. Vancouver, *Voyage*, 3:475.

21. [Barrow], *Mutiny*, chapter 2.

22. Morrison, *Journal*, 83, 84.

23. The log of the *Topaz* is in the collections of the Nantucket Historical Society and has been microfilmed. On p. 12 of his report—see n26 below—Pipon wrote, "By all accounts we could collect from Old John Adams, they have been

upon the Isle about 25 years, but it is impossible to ascertain with certainty the date of their arrival." He goes on to describe Adams's "Journal" as primarily an account book. (Today's Pitcairners celebrate "Bounty Day," 23 January 1790, the estimated date the beached *Bounty* was burned.)

24. Folger, "Log of the *Topaz*," 6 February 1808 entry. Folger's entries are in terms of the nautical day. In addition, he did not repeat a day when the *Topaz* crossed the 180° meridian on 19 November 1807.

25. Because in 1805 the log books of British navy vessels were kept in terms of the civil day, which begins at midnight, the dates given here are in those terms. For much of this early history, including Folger's earlier communications to the Admiralty, see Barrow, *Mutiny*, chapter 8.

26. Pipon, "Narrative." At least three copies of Captain Pipon's September 1814 report exist; all are similar in wording. Two of them are part of ser. 71, "Papers Concerning the Discovery of Pitcairn Island and the Mutineers of HMS *Bounty*." A third one is in the Council for World Mission Archives: Home, Incoming correspondence (1813–21), box 3, folder no. 5, 1817 [*sic*]. My transcription is from ser. 71.06, frame CY 3011/363 of the Sir Joseph Banks Electronic Archive, State Library of New South Wales, with permission to quote.

27. In *Pitcairn*, 4–5, an annotated listing of the vessels stopping by the island, Herbert Ford writes that the islanders were still keeping the Eastern date on 18 October 1817, when Caleb Reynolds, master of the ship *Sultan*, visited the settlement. Ford adds that Captain Reynolds pointed out the day's difference at that time, and as a result "[t]he Pitcairners promptly change[d] their day of worship." Of the several accounts of this stopover, including Reynolds's published and unpublished papers in the possession of the author Emily Reynolds Baker, his descendant, there is no mention at all of a day's difference between the westward-sailing *Sultan* and those living on the island—a difference that surely would have been remarked upon. Without doubt, the Pitcairners had already changed their reckoning to conform to the American dating. (I am indebted to Mrs. Baker for sharing her ancestor's records with me.)

Over the next decades the Pitcairners encountered numerous dating adjustments: their 1831 removal to Tahiti and the subsequent return in two waves, their 1856 removal to Norfolk Island followed by the 1859 return of some to Pitcairn, and then, in 1886, a switch of their day of worship, from Sunday to Saturday. Interestingly, not all of the five date line crossings are mentioned in the relevant contemporary literature.

28. One of the earlier confusions involves Nathaniel Bowditch, who, in 1796, was supercargo on the *Astrea*, sailing from Salem, Massachusetts, to Manila. In his journal this soon-to-be-famous navigator garbles dates.

29. Shillibeer, *Narrative*, 96–97. Shillibeer ascribes this fact to Sir Thomas Staines. Shillibeer's further comment linking the one-day difference to the *Bounty* sailing eastward around the Cape of Good Hope and on to Pitcairn Island implies that the Pitcairners never lost their daily reckoning during eighteen years of isolation. This is an untenable view, given the loss of the ship's arrival date, as stated in the independent accounts of Captains Folger and Pipon.

30. Ibid., 82; see engraving inserted between pp. 92 and 93. Noted first in

Hayes, *Captain from Nantucket,* 52. In his *Voyages aux îles du Grand océan,* first published in French in 1837, the Flemish trader J. A. Moerenhout identified this Pitcairner as "Mardi-Octobre Christian," with whom he conversed on 20 January 1829. Yet another shift in name day! For a translation of Moerenhout's remarks, see *Travels to the Islands of the Pacific Ocean,* 1:25, 439.

31. [Barrow], "Porter's Cruize," 374–83.

32. Ibid., 381.

33. There is actually no way to make Barrow's summary of the *Topaz*'s journey and Captain Folger's encounter with the Pitcairners internally consistent.

34. Lovett, *History,* 132–34.

35. For a superb summary of this early era, including maps, see Gunson, *Messengers of Grace,* 11–28.

36. J. Rodgerson and G. Stallworthy, 25 August 1835, to directors, LMS. See also Gunson, *Messengers of Grace,* 125, which describes an abortive shift to Western dating in 1822 on one of the leeward group in the Society Islands, thus underlining the sensitivity in matters of independent action. Note that the footnote "from the east" should be changed to "via the eastward route."

37. Newbury, *Tahiti Nui,* 128.

38. Lovett, *History,* 346–47. France initiated its annexation of the Austral (Tubuai) Islands in 1880, an integration completed by the end of the decade.

39. As translated in Selga, "Philippines," 10, which includes a discussion of the effect of the shift on the region's ecclesiastical calendar and on commercial and financial documents.

40. Ibid., 11–13. For examples of the erroneous American publications, see "The International Date Line," 85; "Where Does Sunday Begin?" 137; and Collins, *International Date Line,* a so-called revised edition (1891) of a brochure first published in 1881.

41. The Admiralty order directing the change was issued in 1805; American naval vessels did not make the change until 1848—see chapter 7, n2.

42. An important survey having citations to the original documents is Malloy, *"Boston Men."* Its existence made it possible to examine efficiently nearly two dozen surviving logs in the collections of the Phillips Library, Peabody Essex Museum, Salem, Mass.

43. Two pre-nineteenth-century logs are those of the *Columbia* (1790–93) and the *Union* (1794–96), both kept by John Boit, both vessels sailing out of Boston. The dating was not changed when they were at Canton, and both logs recorded the loss of one complete day on arriving home.

44. [Fitzroy], *Narrative of the Surveying Voyages,* 2:512; also Darwin, *Narrative of the Surveying Voyages,* 3:480, 483. Ross, *Voyage of Discovery,* 2:131–32. For another contemporary shift, see Belcher, n46 below.

45. For the text of the 1824 Convention between the United States and Russia, see Davidson, *Alaska Boundary,* 75–78; for a facsimile of the original of the 1825 Convention between Great Britain and Russia, see Huculak, *When Russia Was in America.*

46. Belcher, *Narrative of a Voyage,* 1:98. Arriving in Tahiti on 4 April 1840, Belcher wrote, "I had forgotten the Tahitians were one day in advance of us, the

missionaries having brought their time from the westward [direction]; we were therefore obliged to change our Saturday for Sunday" (1:386).

47. A reconstruction of the Russian settlement forms part of Fort Ross State Historic Park.

48. Veniaminov, *Journals*, 202, 206. The calendar date entries are Old Style.

49. From the extensive scholarly studies of Russia in Alaska and California, I used sections from Essig, "Russian Settlement," 5–25; Davidson, *Alaska Boundary*, 32–34; and Huculak, *When Russia Was in America*, 11–13.

50. Miller, *Alaska Treaty*, 1.

51. Ibid., 1, 130; U.S. President, *Russian America*.

52. Davidson, "International Date Line"; the underscoring has been dropped. Elsewhere in a note (6 December 1898) Davidson quotes from his source, Reverend Sebastian Dabovitch, going on to remark that the latter described the actual change backward. Using Davidson's date for the change, the Orthodox Church held services on Sunday, 20 December 1870 (O.S.), and the next day held Sunday services again—calling this day Sunday, 1 January 1871 (N.S.)—thereby making their days consistent with Western dating, all the while retaining the Julian reckoning.

53. American stories highlighting circumnavigation and/or crossing the antimeridian include Edgar Allan Poe's 1841 "A Succession of Sundays" ("Three Sundays in a Week," 1850), Bret Harte's "The Lost Galleon" (1867), and Mark Twain's *Following the Equator: A Journey around the World* (1897). A bit surprisingly, the American journalist Nelly Bly does not mention gaining a day in the accounts of her 1889–90 race to best Phileas Fogg's time. Of course, like Fogg's, her elapsed time was being monitored at home, in this case at the offices of the *New York World*.

54. The ordinance's text is in Hayden, "Present Status," G6. Howse, *Greenwich Time*, 162, mistakenly describes it as requiring the use throughout the Fiji Islands of a so-called antipodean time; the ordinance is actually cited as "Uniform Date Ordinance, 1879."

55. Davidson identifies the businessman as J. Wrightman and notes the approval of religious groups; see "Notes," Davidson Papers. The idea of diplomatic flattery on the part of the Samoan king comes from Leigh-Browne's summary article, "International Date Line," 306; however, it lacks citations.

56. The author Robert Louis Stevenson lived at Vailima in the hills above Apia and died there in 1894. His mother, whose account of the change in date is often cited, additionally noted that the ship she was taking to Australia had not shifted its dating, her 1 January being the vessel's 2 January 1893; Stevenson, *Letters from Samoa*, 191, 260–61.

57. See *Nature*, "Greenwich Date," 68, 105. Beginning on 9 May 1878 with a short note by the engineer and electrician (Josiah) Latimer Clark and extending through 30 January, the editors of *Nature* printed a half dozen letters under the title "Time and Longitude." This body of correspondence highlights the vagaries of changing the date near the Greenwich anti-meridian, for several anecdotes garble the shift in dates; see also n40 above. In his letter to the American Geographical Society (see n59 below), Davidson cited an 1894 encyclopedia with

an erroneous date line, which had been copied from an 1880 authority; he also noted that "the two best German maps of 1897 gave the line differently."

58. G. Seneca Jones to Representative S. G. Hilborn, 15 May 1898.

59. Davidson, correspondence with the Chief Hydrographic Officer, March–April 1899. In his 3 April cover letter Davidson provides some details regarding the operations of the commercial steamers. He also summarizes his initial interest, his studies, and interactions with the Hydrographic Office in a 6 August 1899 letter to the American Geographical Society; quoted in Strong, "Universal World Time," 483.

60. The chart with Davidson's hand-drawn depiction of the date line is listed as no. 7, "International Dateline. 1899," in Heynen, *USHO Manuscript Charts.*

61. The publication date of the August pilot chart was 25 July 1899. A similar chartlet was printed on the "September Pilot Chart of the North Atlantic Ocean," published 1 September 1899.

62. See Dick, *Sky and Ocean Joined*, 323–29.

63. Very likely Harkness's information came from the 1892 edition of *Stieler's Handatlas*, for a copy of this particular edition, stamped "Nautical Almanac Office Washington, Sep 11 1892," is in the U.S. Naval Observatory Library, and the sweep of the astronomer's date line is identical to the one depicted in this edition.

64. Davidson had focused on those island groups that had regular steamship service, and the Cooks were far from the usual routes. Campbell, *Impressions of Tongareva*, 116–17, 119, 125, 134; Kloosterman, *Discoverers*, 60.

65. Downing, "Where the Day Changes," 176–78.

66. For the false islands, see Stommel, *Lost Islands*, 16–20. In the case of Wrangell Island, Davidson drew his date line west of the island, which extends from 176°W to 179°W, while later maps show the date line bisecting it. For the subsequent history of claims and resolution, see Rudmose-Brown, "Wrangel [*sic*] Island," 23:801.

67. U.K., Admiralty, Hydrographic Department, "Notes," 385–88. The original of the map (with the date lines in different colors) and the associated notes are in the archives of the UKHO, File H6948/1921.

68. An edited version of the map, but without attribution, is in Leigh-Browne, "International Date Line," 303. In its public information program, the U.S. Naval Observatory displays a colorized version of the map as printed in the *New Zealand Journal of Science and Technology.*

69. An annotated copy of the article is in the Davidson Papers. Apparently the *Examiner* is the only place where Davidson's lecture on the subject was ever published.

70. U.K., Admiralty, Hydrographic Department, "Notes," 386.

Chapter 2

1. Struve, "Du premier méridien"; this is a translation of the original Russian text.

2. Each of these imaginary lines begins at an observer's south horizon, passes through his locale's zenith, and terminates at his north horizon. Many such

meridian arcs are defined precisely by a constrained telescope—a transit or transit circle—that sweeps only a very narrow north–south strip of the observer's sky. At major astronomical observatories the position of its transit instrument is often used as the starting point for a series of numbered north–south lines—those on maps and charts—which represent a particular region's longitudes. Obviously, one set of map longitudes can be included with another's if the constant relating the two sets is known. As already pointed out, the location of any zero of longitude is totally arbitrary; latitudes, the second parameter for defining a location, are figured north or south of the equator.

3. Jackson, "Chartography," 181–82. See also U.S. Congress, House, *American Prime Meridian*, for reasons both in favor of and against adopting a national meridian.

4. F. G. W. Struve, *Arc du méridien*.

5. F. G. W. Struve, *Expédition chronométrique . . . entre Poulkova et Altona*; and F. G. W. Struve and Struve, *Expédition chronométrique . . . entre Altona et Greenwich*. These two expeditions, which are magnificently detailed in the reports, were the model for all U.S. Coast Survey longitude expeditions undertaken prior to the Civil War.

6. James, *On the Rectangular Tangential Projection*, 3. Brown, *Story of Maps*, 283, terms James's proposal a Greenwich-based universal-time proposition; however, this is a misreading of Wheeler, *Report*, 30.

7. Wheeler, who in the 1880s examined "specimens of the extended general topographical map series of Europe," lists fourteen separate meridians; Wheeler, *Report*, 30.

8. Thanks to the map historian Ian Mumford for bringing to my attention both the James citation and the subsequent history of the British mapping program.

9. The Convention of the Metre was signed on 20 May 1875 by plenipotentiaries of seventeen countries.

10. "La question de l'unification des méridiens ne dépend d'aucune considération d'économie politique, elle intéresse uniquement le monde savant"; Struve, "Du premier méridien," 49; my translation of the French translation.

11. Thrower, *Maps*, 60–62; Brown, *Story of Maps*, chapter 3; and Haag, *Geschichte*, 1–111.

12. Struve, "Du premier méridien," 58.

13. Wheeler, *Report*, 30–31.

14. In Struve, "Du premier méridien," 60, given as "incomplet et peu précis, quoique cher."

15. Ibid.

16. Obviously France's Paris meridian and Spain's Madrid and Cádiz meridians also bisect both continents.

17. The navigator and U.S. Naval Observatory superintendent Matthew Fontaine Maury urged the selection of an initial meridian "so far west that it would not enter the Atlantic ocean or Gulf of Mexico" so that there would be no east and west reckoning of longitudes by American merchant vessels; U.S. Congress, House, *American Prime Meridian*, 18–20, 25–32. In the end, the United States selected Greenwich for navigational charts and the Naval Observatory's Washington meridian for geographical and astronomical purposes.

18. In his discussion of the so-called American method for determining longitude differences, the astronomer Edward S. Holden noted that the U.S. Coast Survey's first determination (1866) of the difference in longitude between Washington and Greenwich via the transatlantic cable was in error "due to the application of a personal equation with the wrong sign"; in Holden, *Memorials*, 246–47.

19. Two other exact-hour meridians satisfy Struve's criterion: 165°E and 165°W.

20. Struve, "On Universal Time," 170. I will discuss the proposal for a universal or cosmopolitan time later.

21. Struve, "Du premier méridien," 63–64.

22. Ibid., 64.

23. IGC 1, *Compte-rendu*, 1:206–7.

24. Ibid., 1:176, 183, 184, 206–9, 381; 2:234, 254–57.

25. Ibid., 1:176.

26. Ibid., 1:206–8. Curiously, several participants described Struve's selection of the Greenwich anti-meridian as passing through the Bering Strait and separating Asia from America. Actually, the difference between the two is quite significant: eleven degrees of longitude, or forty-four minutes in time.

27. Ibid., 2:254–55.

28. Ibid., 255–56. I believe the unidentified participant strongly opposed to the congress's deliberations was the French hydrographer Adrien Germain, whose views were published in 1875 (see n30 below).

29. Ibid., 257.

30. Struve, "Du premier méridien," 46–64; Germain, "Le premier méridien," 504–21.

31. Ibid., 507, 508. Germain's article was printed during the diplomatic sessions that led to the signing of the Convention of the Metre (20 May 1875). Great Britain did not sign the convention.

32. Ibid., 511–12. *The International Code of Signals*, drafted by the British Board of Trade, was first published in 1857. It and subsequent revisions contained a section of universal signals to be used by all nations. The code was adopted by most maritime countries.

33. Ibid., 516–21.

34. Fleming, "Longitude and Time-Reckoning," 149. Unaccountably, Fleming also concluded that Germain—who "seems to think, for his opinions are not positively expressed"—*favored* the British Greenwich–French meter quid pro quo, this a canard repeated by at least two twentieth-century British historians of timekeeping (Humphry Smith and Derek Howse).

35. Charles Ploix, "Utilié d'une entente," 1:61.

36. IGC 2, *Compte rendu des séances*, 1:34.

37. Ibid., 33–35, 60–61; 2:38, 401.

38. Of the dozen and a half active coastal surveys reported in the 1880s, all but one group—consisting of those conducted by the U.S. Coast and Geodetic Survey—were carried out within navy departments or marine ministries; see "Marine Surveys" in Wheeler, *Report*, 497–98.

39. The geologist called his choice the meridian of Ptolemy adopted by Mercator. Struve's earlier suggestion of a meridian two hours from Greenwich (32°20′ west of Paris) is of course similar to this Azores meridian with regard to advantage of location.

40. IGC 2, *Compte rendu des séances*, 1:26–27.

41. Ibid., 29.

42. Ibid., 30. Another reason proffered was the creation of a logical link to the modern era of chronology based on Christianity. This particular choice did not receive any great attention at this congress. However, the question of Jerusalem as the common initial meridian was discussed again at the Fourth (1889) and the Fifth (1891) IGC, owing to the interest of some members of the Academy of Sciences of Bologna. It was also brought to the attention of the International Telegraph Conference meeting in Paris in 1890.

43. U.S. Department of Commerce, NBS, *International Bureau of Weights and Measures*, 225–31.

44. In a report, "The International Congress and Exhibition of Geography," the editors of *Nature* noted that the adoption of a first meridian had been "very much discussed." However, they cited only Struve's current proposal, incorrectly terming it "Greenwich."

45. IGC 2, *Compte rendu des séances*, 1:30. While one could suggest that the French astronomer was deliberately deflecting any movement toward unification by setting conditions requiring enormous financial and technical resources and literally decades of effort, much of what he was proposing became the long-term focus of specialists in geodesy and topographic surveying.

46. IGC 2, *Compte rendu des séances*, 2:401.

47. Among the latter was the International Congress of Commercial Geography, which met in Paris in 1878 and in Brussels the following year.

48. De Claparède, "Henry Bouthillier de Beaumont."

49. The date is from de Beaumont's later writings. The summary, "Le méridien unique," has the event as 17 January 1877, apparently a typo. This meridian is similar to Struve's in that it bisects a nearly uninhabited region of a continent—in de Beaumont's case, Alaska's Cape Prince of Wales.

50. This meridian lies 12° east of Greenwich and passes through Norway, Sweden, Denmark, Germany, Austria, and Italy.

51. De Beaumont, "Choix d'un méridien initial," and "Note [d'un méridien initial]." His fifteen-page pamphlet, with map, *Choix d'un méridien initial unique*, is the one cited by later historians. "A New First Meridian," *Times* (London), 16 January 1879, 6, and ibid., *Times Weekly Edition*, 17 January 1879, 9; "Many proposals," *Nature*; and "One Evidence of National Weakness," *New York Times*, 10 February 1879, 4. These editors ignored de Beaumont's use of the Bering Strait anti-meridian lying in Europe to reckon longitudes; see de Beaumont, "Note [d'un méridien initial]," map facing p. 208.

52. De Beaumont's writings in this area are extensive and continued to at least 1895, long after both the idea of a neutral meridian and his Bering Strait proposal had been rejected by European geodesists and others. In my judgment, his efforts have been greatly overemphasized in English-language histories of

the era. Many others also proposed meridians; see the citations to works of astronomers and geographers given in the official reports of the International Geographical Congresses and in Wheeler, *Report*, 32–33.

53. During this era, the value of passing such resolutions by international organizations that lacked power to effect change was challenged by several authorities. See, for example, the remarks by Wilhelm Foerster, director of the Berlin Observatory, reported in *Proceedings of the Royal Geographical Society* 13 (1891): 733–34, after the Fifth IGC in Berne.

Chapter 3

1. Dowd, "Origin and Early History."

2. Starting in 1850 the initial meridian for America's geographical mapping and astronomy passed through the axis of a U.S. Naval Observatory transit telescope at its then-current site in Foggy Bottom—see Dick, *Sky and Ocean Joined*, 124–27; the north-to-south-running borders of several western states are linked to it. Congress repealed the American prime meridian statute in 1912.

3. Dowd, *System of National Time*. Dowd's map has been reproduced widely: in color on the cover of *Railroad History* 148 (Spring 1983); in Stephens, *Inventing Standard Time*; and in Bartky, *Selling the True Time*, 99.

4. For a history of Dowd's efforts, including a detailed comparison of the educator's proposals and what was actually adopted in 1883, see Bartky, "Invention of Railroad Time," "Adoption of Standard Time," and *Selling the True Time*, 97, 100.

5. Soon after a plaque honoring Sandford Fleming's contributions to Standard Time was erected in Canada, the Harvard astronomer and science writer R. Newton Mayall wrote "The Inventor of Standard Time." Using Fleming's published materials, Mayall debunked many of the priority claims inscribed on the memorial. At the same time, however, Mayall exaggerated Dowd's contribution to the railroads' 1883 adoption of Standard Railway Time.

6. Despite a plethora of articles and books, no satisfactory biography of this important nineteenth-century Canadian exists. The authorized biography—Burpee, *Sandford Fleming*—cannot be considered an objective study; a work by the creative writer Clark Blaise, *Time Lord*, is replete with errors of fact and misstatements and is simply not a historical analysis. In contrast, the late Mario Creet's analysis of one facet of Fleming's career, "Sandford Fleming and Universal Time," is well documented and worthy of close attention.

7. The anecdote is in Fleming, *Terrestrial Time*, 5–6, and *Uniform Non-Local Time*, 5–6. Fleming's wife and children were staying in a resort town near Londonderry, so the unexpected delay meant that his family lost a day of sightseeing in the north of Ireland.

8. For the first-ever analysis of the missed-train incident and a detailed examination of Fleming's writings on timekeeping in 1876–79, see Bartky, "Sandford Fleming's First Essays."

9. Howse, *Greenwich Time and the Longitude*, 129. However, nothing in Dowd's correspondence in the Fleming Papers nor in Fleming's extensive col-

lection of time-related pamphlets at the Library and Archives Canada indicates that he was aware of Dowd's efforts prior to 1881.

10. Burpee probably based his dating of *Terrestrial Time* on the railway engineer's use of a literary device introducing the incident—"A few weeks ago he (the writer) . . ."—and Fleming's own dating in his "Bibliography," 33, which Burpee used to create his subsequent listing; see Burpee to Fleming, 27 March 1907, Fleming Papers, vol. 7, folder 44.

11. Fleming published three nearly identical pamphlets: *Terrestrial Time* (March 1878); *Temps terrestre: Mémoire* (August 1878), a translation of the first one commissioned by Fleming for a presentation in Paris that never came to pass; and *Uniform Non-local Time (Terrestrial Time)*, partially printed in North America, in which the last five pages of *Terrestrial Time* have been discarded, and which became available around November 1878. For the detailed bibliographic analysis, a portion of which is due to Andrew S. Cook, India Records Office, British Library, see Bartky, "Fleming's First Essays," n8 above. In *Selling the True Time*, 148, I indicated that Fleming's first pamphlet was published in 1876; the research here corrects this misdating.

12. In his subsequent work, "Time-Reckoning," 98, Fleming had his traveler continue to San Francisco via Detroit and Chicago.

13. Dowd, *System of National Time*, 3; Fleming, *Terrestrial Time*, 2.

14. In the first section Fleming noted what he termed the "natural measures of time"—the year, month, and day—citing their variety: solar, sidereal, mean solar, and astronomical times. In the second section he listed a few of the ways various peoples have subdivided the day, adding to the list timekeeping on board ship and the system of sailors' duty watches; see Fleming, *Terrestrial Time*, 8–12.

15. Fleming used many terms in this first published work: "terrestrial time," "non-local time," "common time," and "universal time"; later he would use "cosmopolitan time" and "cosmic time," as well as suggesting "absolute time" and "all world time."

16. In subsequent iterations, Fleming assigned the letters of the alphabet to different meridians; see n36 below.

17. Linking the terrestrial-time meridians to Greenwich makes them identical to the hour-difference ones articulated by Otto Struve in 1870.

18. As an example of his notation, 12:02 P.M. along the Greenwich meridian would be written as G.02, midnight Greenwich time would be T.00, and 6:23 A.M. at Greenwich would be A.23.

19. Fleming, *Terrestrial Time*, 20. Of Fleming's nearly three dozen clock dials depicting terrestrial time, only one can be used on an ordinary timekeeper—see his figure 5, p. 28—and this one is marred by his attempt to incorporate ships' so-called night watches on the dial plate. Moreover, because of the twenty-four regions of the globe he describes, different dials would be required on each timekeeper. Furthermore, Fleming's layout of the dial plate required that the hour hand be the longer one, a new configuration running completely counter to all current clocks and watches. On 10 March 1879 Fleming filed for a U.S. patent on his watch, listing six claims. Five were rejected as being prior art, and the last was rejected because it was written improperly; see Fleming Papers, vol. 102, folder 33.

20. Fleming, *Terrestrial Time*, 17, 18. A precursor letter-hour dial exists. In an article titled "Time for the Continent," X. Sentrick—certainly a pseudonym—proposed a single railroad time, that of New York City, to be distinguished by a set of twenty-four letters running from A to X. The illustration accompanying the proposal is of a twelve-hour timekeeper whose dial has three concentric rings: one with the usual twelve numbers for displaying local time and two rings of letters, A–L and M–X. Letters A and M are linked to 12; thus, a railroad time of M:15 would translate as 12:15 P.M. local time, while A:20 would be 12:20 A.M. local time. This particular way of displaying railroad time and local time simultaneously is quite similar to Fleming's later, more complicated displays.

21. Writing in July 1879, Astronomer Royal George Airy predicted correctly that Fleming's clock faces would never be adopted, and if timekeepers having such dials "were exposed in shops, I do not suppose that one would be sold"; quoted in Fleming, *Universal or Cosmic Time*, 34. Beginning on 13 November 1884 and continuing through 5 February 1885, the editors of *Nature* published a series, "Our Future Watches and Clocks," giving the pros and cons of various ways to adapt timekeepers to a set of twenty-four numerals, thereby documenting the demise of lettered hour dials.

22. Of the fifteen station entries in the Cork–London double timetable, Fleming misconverted all five Irish times by making them twenty minutes ahead of Greenwich instead of twenty-five minutes behind and then misconverted the Greenwich times of Gloucester, Swindon, and Oxford—a total of eight errors. The London–Pacific Coast pair contains five conversion errors. See also n35 below.

23. Fleming, *Terrestrial Time*, 23.

24. Dowd of course needed to keep the two times separate in order to market his gazetteer. The idea of one time for all public purposes had already been advanced by the American astronomer Samuel Langley in his 1870 uniform-time proposal (see Bartky, *Selling the True Time*, 96).

25. Illustrated by a small diagram in Fleming, *Terrestrial Time*, 34.

26. Ibid., 36.

27. Fleming's biographers highlight what they claim was shabby treatment at the August 1878 annual meeting of the BAAS in Dublin, where Fleming was unable to present his scheduled communication; I do not agree with this interpretation, which is based on Fleming's (draft) letter of complaint and remarks made years later. Fleming also wanted to present his views in September 1878 during the Exposition Universelle in Paris; however, he failed to find a sponsor. For an analysis of these two incidents, see Bartky, "Fleming's First Essays," n8.

28. Fleming, *Terrestrial Time*, 22.

29. Fleming, "Time-Reckoning" and "Longitude and Time-Reckoning." According to local newspapers, Daniel Wilson, president of the Canadian Institute, read the first paper for him at a regular meeting on 25 January 1879, and Fleming himself presented the second one on 8 February. A historical Canadian painting by Rex Woods, "The Birth of Standard Time," commemorates the latter event. Currently the original hangs in the lobby of the Rogers Communications building, 1 Mount Pleasant Road, Toronto.

30. In this work Fleming neglected to alter his illustrations so as to make them consistent with his new lettered meridians, certainly leading to great confusion for any contemporary reader. In mid-September 1878 Fleming learned that the AMS had been considering nearly identical timekeeping issues for several years, so he may have rushed into print without correcting his pamphlets' tables; see Bartky, "Fleming's First Essays."

31. Fleming wrote that he saw both Otto Struve's and Adrien Germain's articles (discussed in chapter 2) published in the 1875 *Bulletin de la Société Geographique*; see "Longitude and Time-Reckoning," 149. He also quoted verbatim "A New First Meridian" from the *Times Weekly Edition*, in which summaries of the initial meridian discussions at the 1871 and 1875 International Geographical Congresses were given; see "Longitude and Time-Reckoning," 147–48.

32. IGC 2, *Compte rendu des séances*, 1:30.

33. See chapter 1, especially n57, for examples.

34. Although in use by much of the general public and railways, not until 1880 did Parliament enact legislation making these two times the kingdom's official (legal) ones.

35. Fleming did not correct the earlier cosmopolitan time errors in his London–Pacific Coast timetable pairs; see n22 above. Moreover, he neglected to shift the letter designations for his cosmopolitan time illustrations, so all twenty-four of them are wrong.

36. In "Time-Reckoning," Fleming shifted the sequence of letters so as to link Y to his zero meridian; the Greenwich meridian, formerly G, became M. Later (in August 1881) Fleming assigned "Z" to his initial meridian. When time zones are lettered today, Z is the Greenwich meridian; the alphabetical sequence of letters is split and runs both east–west and west–east to conform with the numerical longitudes currently in use. Also, J is dropped, and an additional letter is used to distinguish the different days in the zone bisected by the International Date Line, producing a total of twenty-five letters for the zones.

37. "Longitude and Time-Reckoning," facing p. 146. Note in the explanatory footnote on p. 146 that the time difference for meridians to the west should be "one hour *earlier*," not "later."

38. Struve, "Du premier méridien"; Fleming, "Longitude and Time-Reckoning," 141, 143. Howse, *Greenwich Time and the Longitude*, 131, quoting an 1883 source, gives annual sales of the *Nautical Almanac* as 20,000, with 3,000 for the *Connaissance des temps*.

39. Fleming, "Longitude and Time-Reckoning," 142. As usual, some arithmetic errors mar Fleming's table; however, they do not affect his argument.

40. Ibid., 146.

41. See the discussion in chapter 2 and the citations listed in nn1 and 30.

42. It bears repeating here that de Beaumont's initial idea was to project his Bering Strait meridian over the poles, reaching into the European continent. There he would begin his east and west longitudes.

Chapter 4

1. International Meteorological Congress, *Berichte des Professor Dr. C. Bruhns*, 17–24; *Procès-verbeaux*, 16–18; *Rapports sur les questions du programme*, 195–96.

2. For events and details of the process, see Bartky, "Adoption of Standard Time," 25–56; and *Selling the True Time*, chapters 7 and 10–12.

3. Abbe, "Report." In *Selling the True Time*, 206–7, I noted the unnecessarily long time taken—four years—to prepare the report. For a speculation regarding its completion, see n16 below.

4. Abbe, "Report," 28, app. V. Very likely this opinion was inserted as a result of the recently inaugurated collaboration between Fleming and the AMS's Barnard and Abbe.

5. The Marquis of Lorne was the eldest son and heir of the Duke of Argyle; his wife was Princess Louise, fourth daughter of Queen Victoria. The Canadian Institute, now the Royal Canadian Institute, a scientific organization in Toronto chartered in 1851, had been founded by Sandford Fleming and others.

6. The memorial and subsequent official correspondence can be found in Fleming, *Universal or Cosmic Time*, Supplementary papers, 28–44.

7. Creet, "Sandford Fleming and Universal Time," 74, calls the step taken by the Canadian Institute both "unusual" and "unprecedented," but does not provide any comparisons. It seems certain that Fleming had a strong hand in urging this course.

8. Airy, as given in Fleming, *Universal or Cosmic Time*, 33; his allusion to "some personal influence" remains a mystery. In August of the following year the Statutes (Definition of Time) Act made Dublin the time of Ireland and Greenwich the time of Great Britain in all official documents.

9. See Wilkens, "History," 56, regarding the almanac's poor reliability in the early 1800s.

10. Airy, quoted in Fleming, *Universal or Cosmic Time*, 33. Airy also indicated his objection to counting longitudes in one direction from 0° to 360°.

11. Ibid., 34; Airy's response to the secretary of state for the colonies is dated 18 June 1879.

12. The usual title is *Papers on Time-Reckoning and the Selection of a Prime Meridian to be Common to all Nations*. Some printings may differ slightly in title.

13. The Colonial Office distributed copies in Great Britain.

14. Fleming, *Universal or Cosmic Time*, 30, 35–38; Royal Astronomical Society, *Minutes*, also noted in Sadler and Wilkens, "Astronomical Background," 197; Royal Geographical Society, "Sir J. H. Lefroy's Report," 19 November 1879; I am indebted to Francis Herbert of the Royal Geographical Society for bringing this review to my attention. The Royal Society report, dated 6 November 1879, is quoted in Fleming, *Universal or Cosmic Time*, Supplementary papers, 39.

15. Secretary of state for the colonies to governor-general of Canada, 15 October 1879, as quoted in ibid., 31–32.

16. Barnard (1809–1889) was a distinguished educator and former astronomer who was also the president of Columbia College. Fleming's interaction with the AMS president in the fall of 1878 may possibly have led to Barnard's concern over being scooped by the Canadian and thus to the prompt completion of the AMS's own report. For the probable impact on Fleming, see n30 in chapter 3.

17. Portions of this correspondence are in Abbe, *Professor Abbe*, 145–48. In proposing that the government of Canada act in the matter, Abbe was obviously unaware that the imperial government's official position precluded such an approach.

18. Fleming, *Universal or Cosmic Time*, Supplementary papers, 44–48, 52–55. In addition, in 1881 a Spanish navy officer opposed to the Geographical Society of Madrid's recommendation of a meridian passing through Ferro translated Fleming's 1879 papers, which were then published by the Ministry of Marine; ibid., 49–51. He also advocated Fleming's ideas in 1884 at the International Meridian Conference.

19. On 22 May 1880 Fleming, whose engineering leadership was being heavily criticized, was either "relieved . . . of his duties" (Green, *Chief Engineer*, 104) or "resigned" as engineer-in-chief (Berton, *Impossible Railway*, chronology, 540).

20. See Creet, "Sandford Fleming and Universal Time," 77.

21. The paper was published in *Transactions of the ASCE* 10 (December 1881): 387–92. The chart accompanying it, which displays the globe partitioned by cosmopolitan time meridians, shows evidence of hasty preparation as well as a change in notation: instead of twenty-four, twenty-*five* lettered meridians are indicated, and the zero meridian is labeled Y, not Z. For some details of Fleming's efforts within the ASCE, see Creet, "Sandford Fleming and Universal Time," 78–80, 85, 87–88.

22. See Bartky, *Selling the True Time*, 148–49.

23. IGC 3; Wheeler, *Report*, 270.

24. Fleming, "Adoption d'un maître méridien," in IGC 3, 2:10–17; later translated and published as "On the Regulation of Time." Hazen, "Adoption," IGC 3, 2:18–19; Barnard, "Adoption," IGC 3, 2:7–9.

25. For the American and Canadian resolutions, see Wheeler, *Report*, 24–27.

26. IGC 3, 1:247–48.

27. From "List of Views Issued by Each Group and Not Presented at the General Sessions," IGC 3, 1:392. Wheeler, *Report*, 23, gives a translation extract with a somewhat more positive tenor. Another translation is in Fleming, *Universal or Cosmic Time*, Supplementary papers, 63–64.

28. See IGC 3, 2:xv–xviii, for the circular letter and summaries of the ten replies. As mentioned in n14 above, the Royal Geographical Society's formal response to the Italian Geographical Society's 19 July letter came after a review of Fleming's proposal by its president, General Sir J. H. Lefroy.

Chapter 5

1. See Bartky, *Selling the True Time*, 119 and chapter 11, for a detailed discussion of the adoption process in the United States.

2. Creet, "Sandford Fleming and Universal Time," 77–78, argues that the ASCE Special Committee on Standard Time "was to be the main actor in 1882." Considering the other time-related events taking place in the United States throughout 1882, including the congressional debates and subsequent passage of the joint resolution in July, I cannot accept this view.

3. Fleming, "Cosmopolitan Scheme," and Fleming, *Letter*, 3–10.

4. The replies were received in May; Creet, "Sandford Fleming and Universal Time," 79, provides some statistics. The most-often-quoted response is that of Simon Newcomb, superintendent of the U.S. Navy's Nautical Almanac Office, who answered one general question with "A capital plan for use during the millennium" and also "See no more reason for considering Europe in the matter than for considering the inhabitants of the planet Mars." Fleming himself popularized this negative response as a way to denigrate Newcomb's opposition to other timekeeping changes that he desired. Never quoted is the astronomer's precise answer to no. 1 of the questionnaire: Are you in favor of a comprehensive system of Standard Time for North America? "YES," Newcomb responded. As he wrote later in his *Reminiscences*, 225–26, he favored Washington time as the country's standard, which was also the choice of the U.S. Naval Observatory astronomers.

5. From Fleming, *Letter*, 5.

6. At the society's 27 December 1881 meeting, Barnard informed attendees of a private letter from his friend and ally Fleming. In it the Canadian suggested that as a way to recover from the failure to accomplish anything substantial at the Third International Geographical Congress, the president of the United States be approached to "intimate to the Italian Government the meeting of such a convention at Washington would be agreeable to . . . [the American] government." J. E. Hilgard, superintendent of the USC&GS, then remarked that such an action "would amount to an invitation, which the President would probably not feel authorized to extend without some previous action of Congress. . . ." After further discussion, the society began preparing a memorial to Congress; see *PAMS* 2 (1882): 305–6.

7. A copy of the memorial, with Fleming's signature on it, is in the Fleming Papers, vol. 14, folder 95; its cover letter, from Thomas Egleston, secretary of the AMS, is dated 6 February 1882. Reprinted in *PAMS* 2 (1882): 306–8.

8. See Bartky, *Selling the True Time*, chapter 10, for the details of Abbe's efforts and the resulting conflict between the U.S. Signal Service and the U.S. Naval Observatory.

9. U.S. Department of State, Secretary, Circular Letter, 23 October 1882.

10. His certificate of election to AAAS membership is in the Fleming Papers, vol. 1, folder 8. Fleming's later remarks to the ASCE (18 January 1883) and his pocket diary document his participation in the August 1882 discussions at the AAAS's annual meeting in Montreal, even though his *Letter to the President of*

the AAAS carries an apology for not attending the sessions. Creet claims that as a result of Fleming's input, the AAAS established its own time reform committee, which is not correct. As already noted, its Committee on Standard Time was established in 1880; after passage of the joint resolution by Congress, the AAAS officers shifted both timekeeping and prime meridian matters to its new committee; see Bartky, *Selling the True Time*, 254n14.

11. Fleming, *Letter*, 6. At the next annual meeting the AAAS discharged both time committees. The Committee on Standard Time reported that "the Railway companies had taken such actions as left little for the Committee to do," while the Committee on a Prime Meridian and an International Standard of Time filed no report. See AAAS, *Proceedings of the Thirty-second Meeting*, 481, 482.

12. See Fleming, Chairman's Reports, 18 January 1882, *Proceedings of the ASCE* 8 (1882): 5; 17 May 1882, ibid., 75; and 17 January 1883, ibid., 9 (1883): 3.

13. *Proceedings of the ASCE* 8, 8 March 1882, 8 (1882): 57–58; ibid., 17 May 1882, 8 (1882): 75–76; ibid. 17 January 1883, 9 (1883): 2–4. At its 17 January annual meeting the ASCE adopted resolutions favoring such a convention, with the proviso that other societies be queried as to their participation; see *Proceedings of the ASCE* 9 (1883): 4–5. Fleming approached the AMS in May 1883 with a proposal for cooperation. He was not successful; see Creet, "Sandford Fleming and Universal Time," 80, for details.

14. Bartky, *Selling the True Time*, chapter 11. Additional details are in Bartky, "Adoption of Standard Time."

15. The General Time Convention was not disbanded. Instead, it grew into the industry's technology organization and renamed itself the American Railway Association. Today it is the Association of American Railroads.

16. In an all-too-common response to such major events, a member of the Special Committee on Standard Time gave undue credit to the ASCE for the adoption process: (1) he denigrated the efforts of Charles Dowd; (2) he claimed that the AMS had been concerned entirely with the theoretical side of the matter and had "about abandoned the question when [in June 1881] . . . this matter was brought up before this Society"; (3) he asserted that "the ASCE . . . would have introduced the system into the railways of the country, but a misunderstanding caused delay; and during that time other people took the matter up"; and (4) he even predicted that the upcoming International Meridian Conference would have "little to do except to adopt . . . the plan which is laid out for them by this Society"; see *Proceedings of the ASCE* 10 (1884): 39–40. Unfortunately, this biased, unsupportable view has been promulgated in the ASCE's histories, including its sesquicentennial one. It must be noted here that the members of the AMS were more objective in their reviews of the history of the rail companies' promulgation of Standard Railway Time.

17. U.S. Department of State, Secretary, Circular Letter, 1 December 1883.

18. Curiously, in his summary of events through 1884, Sandford Fleming gives only passing mention to the International Geodetic Association's 1883 conference; *Universal or Cosmic Time*, 8, 10. And in an otherwise quite useful account, Creet ignores the Rome conference in his "Sandford Fleming and

Universal Time," even though he cites Howse's *Greenwich Time and the Discovery of the Longitude*, where the Rome meeting is ranked as one of "two special conferences" (the other being the International Meridian Conference of 1884).

19. IGC 3, *Comunicazioni e Memorie*, 2: XVI–XVII; Geographical Society of Hamburg, *Mittheilungen*; IGA, "Rapport de la Commission," 128. Kirchenpauer's letter was probably sent on 26 November 1882.

20. Levallois, "History"; and Mueller, "125 Years"; the number of states and countries given in the text is from Mueller. In 1886 the Europäische Gradmessung became the Internationale Erdmessung, or International Geodetic Association.

21. IGA, "Rapport de la Commission," 128–30. General Carlos Ibáñez of Spain was president, with secretaries Adolphe Hirsch, director of the Neuchâtel Observatory in Switzerland, and astronomer Theodor von Oppolzer of Austria.

22. Not stated in the report, but probably France.

23. IGA, *Comptes-rendus des séances*, 119–21; IGA, "Rapport," 132–34.

24. IGA, *Extrait des Comptes-rendus*, 14–15. Undoubtedly Hirsch was also reminding his audience of the International Bureau of Weights and Measures at Sèvres, funded by signatories of the Convention of the Metre, and still experiencing growing pains.

25. A summary of Glydén's ideas is part of the record of the International Meridian Conference.

26. To illustrate, 6 A.M. Monday, civil time, is 1800 astronomical time, of the previous day, while civil time's 6 P.M. Monday is the equivalent of astronomical time's 0600 on the same day of the week.

27. Subsequent debates associated with changing the astronomical day are highlighted in chapter 7.

28. For the proposed resolutions, see IGA, *Extrait des Comptes-rendus*, 21–22. The views regarding the introduction of a universal time are in ibid., 34–35. Because of wide interest by non-geodesists, the forty-eight-page French-portion of the sessions was extracted from the *Procès verbaux* and printed separately by the association's Central Bureau as Hirsch and Oppolzer, *Unification des longitudes*; for the Permanent Commission's report see IGA, *Extrait des Comptes-rendus*, 6–22.

29. One of the added resolutions (No. II), which concerned the decimalization of the circle and of time, had been argued since at least the First International Geographical Congress in 1871. Never accepted by a majority of specialists, it would continue as a proposed subject for study at the 1884 International Meridian Conference.

30. IGA, *Extrait des Comptes-rendus*, 47–48.

31. Ibid., 29, 41–42, 44.

32. Resolution VIII, with discussions leading to it, is at ibid., 30, 36–37, 38n, 40, 48; it is significant that the association voted against "La Conférence émet le voeu," substituting the weaker "La Conférence espère"; ibid., 44. The issue of Britain's adhesion to the Convention of the Metre was raised again during the International Meridian Conference.

33. Among the nonrespondents at this time was Great Britain; see F. T. Frelinghuysen to T. Egleston, 20 August 1883, in Fleming Papers, vol. 14, folder 95; presumably France was also a nonrespondent. See also the brief remarks regarding the U.S. invitation made by the American delegate, R. D. Cutts, at the close of the Special Commission's sessions (IGA, *Extrait des Comptes-rendus*, 37–38), and his later ones in Cutts, "Communication."

34. IGA, *Extrait des Comptes-rendus*, 22, 38, 40, 44–45, 48.

35. A tabular summary of the nine IGA resolutions is in chapter 6.

36. Cutts commented that the new American system, called Standard Railway Time, was tied to the Greenwich meridian, but he did not explain its essential linkage to civil time; "Communication," 38.

Chapter 6

1. A copy of the secretary of state's invitation is in U.S. President, *Message . . . Relative to International Meridian Conference*, Annex IV. The date shown there, "December 1, 1884," is a typographical error. In the body of the text is a reference to the "Geographical Congress held at Rome"; it should be the "Geodetic Conference held at Rome."
In 1884 English and French sections of the conference proceedings were printed by Gibson Bros. of Washington, D.C., and bound separately. The English portion carries the title *International Conference held at Washington for the Purpose of Fixing a Prime Meridian and a Universal Day: Protocols of the Proceedings*. Its pagination is entirely different from the official document just cited and used here.

2. *Science*, "International Standard Time," and "Comment and Criticism" (4 January 1884); *Nature*, "The International Conference."

3. U.S. President, *Message . . . Relative to International Meridian Conference*, Annex II. For the conversion to current values, see the process outlined in McCusker, "How Much Is That in Real Money?" with extensions to this century.

4. For the Senate in 1882, see Bartky, *Selling the True Time*, 262n1. The appropriations bill for fiscal year 1885, H.R. 7380, was introduced in the House on 21 June 1884 and passed on 23 June.

5. *Science*, "Comment and Criticism" (27 June 1884).

6. U.S. Department of State, Secretary to Barnard, 17 September 1884. Both of Barnard's resignation letters are dated 22 September 1884 and are included in Barnard, Letter to Secretary of State.

7. U.S. Department of State, Secretary to Meridian Conference Delegates, 24 September 1884.

8. According to Creet, the governor-general of Canada lobbied through the Royal Society to have Fleming included; however, the citation he gives there may not be the source for this statement; "Sandford Fleming and Universal Time," 83. That some atypical activity was occurring during the British selection process can be inferred from the complaint by the Science and Art Department to the secretary of the treasury regarding multiple inputs into the process; see U.K.

Science and Art Department, *Thirty-first Report* [for 1883]: Minutes and Correspondence, 38–39.

9. *Comptes rendus* 96 (1883): 42–43, 109, 135–36, 182–85, 1379–83; 97 (1883): 1234–39; 98 (1884): 407, 453.

10. Caspari, *Rapport.*

11. The fourth resolution incorporated the view that the decimal system be applied to the measurement of angles and time, a long-held French position and one that was passed over at the IGA conference in Rome.

12. U.S. President, *International Conference,* 10–11.

13. Now the Indian Treaty Room in the Old Executive Office Building.

14. U.S. President, *Message . . . Relative to International Meridian Conference,* Resolution I: 17–18.

15. The motion and ensuing comments are at ibid., 22–49.

16. A likely explanation for Haiti's voting consistently with France's position lies in its history. The former Saint Domingue was a French colony but has been the independent country Haiti since 1804; the other country with which it shares the island of Hispaniola (also known as Haiti) is the Dominican Republic. "San Domingo" is an old name for Haiti. I thank copy editor Barbara Norton for bringing this historical and geographical information to my attention.

17. U.S. President, *Message . . . Relative to International Meridian Conference,* Resolution II: 15–17, 18–22, 49–58.

18. Ibid., 51, 54. Curiously, no record was taken of this vote.

19. Ibid., 51–52. Howse, *Greenwich Time and the Longitude,* 137, errs in this regard.

20. Comité international des Poids et Mesures, *Procès-verbaux,* 51–53, 128–37. The run-up to this decision, in which Lieutenant General Strachey participated, is in U.K. Science and Art Department, *Thirty-first Report* [for 1883]: Appendix A, Minutes and Correspondence, 32–42. One discussant, at the Standards Department of the Board of Trade, was critical of those at Rome who had linked adherence to the Convention of the Metre with the adoption of the Greenwich meridian as the common initial meridian, seeing these as two entirely separate issues; ibid., 40.

21. U.S. President, *Message . . . Relative to International Meridian Conference,* 29, 56–57, 57–58.

22. Ibid., Resolution III: 59–77.

23. See Philosophical Society of Washington, "Resolutions," *Bulletin* 6 (1884): 107–9; Hilgard's remarks are at 109. Hilgard was one of the distinguished invited visitors at the Meridian Conference and spoke briefly at one of the later sessions; see U.S. President, *Message . . . Relative to International Meridian Conference,* 99.

24. Ibid., 59. According to Newcomb, *Reminiscences,* 144 and 312, the Russian minister was a brother of Otto Struve.

25. U.S. President, *Message . . . Relative to International Meridian Conference,* Resolution IV: 77–83.

26. Ibid., Resolution V: 84–85, 88–103.

27. Impressed by the concepts outlined by Sandford Fleming in his Canadian

Institute articles, Pastorin had them translated into Spanish. See his "Remarks on a Universal Prime Meridian, by Don Juan Pastorin, Lieut-Commander of the Spanish Navy"; in Fleming, *Universal or Cosmic Time*, 49–51. See also chapter 4, n18.

28. U.S. President, *Message . . . Relative to International Meridian Conference*, 98.

29. In what are obviously transcription errors, at this point the printed document shows the Netherlands as both voting in favor of the resolution and abstaining from the vote; also, Russia is not listed in the tabulation even though all three members of the delegation were present at the session. The earlier, verbal comments of the Russian delegation and the final act indicate that "Russia" should replace "Netherlands" in the listing of countries in favor of the resolution here; see ibid., 59, 86, 102–3, 112. Turkey changed its vote on the day of the last session; see ibid., 113.

30. Ibid., Resolution VI: 50, 103. The Spanish delegate's remark is at p. 52.

31. Ibid., Resolution VII: 103–6. English versions of the resolutions adopted at the Rome conference are probably translations by the USC&GS official delegated to it, and are in Philosophical Society of Washington, "Resolutions," 107–9, and in *Science*, "Resolutions," 2 (1883): 814–15. The decimalization of time and angle was a recurrent theme among some French astronomers and geodesists, who were unsuccessful in convincing colleagues in other countries. A final attempt came via the Bureau des Longitudes in the late 1890s; it too failed. For a useful and absorbing account, see Galison, *Einstein's Clocks*, 163–73.

32. Contemporary summaries of the Washington conference exist, among them *Science*, "Meridian Conference"; Ellis, "Prime Meridian Conference"; and Struve, *Beschlüsse*; an English translation of this excellent work was published in Fleming, *Universal or Cosmic Time*. In his own "Report on the Washington International Conference," Fleming focuses primarily on his individual efforts, as does Janssen in his "Sur le Congrès de Washington" and in an additional summary analysis, "Notice sur le Méridien et l'Heure universels." Howse's brief summary in *Greenwich Time and the Longitude*, 133–43, has some errors of fact and a few unsupportable conclusions.

33. U.S. President, *Annual Message*, 1 December 1884; and *Message . . . Relative to International Meridian Conference*, 4 December 1884. U.S. Department of State, Secretary to Chairman, Senate Committee on Foreign Relations, 5 February 1885.

34. U.S. Congress, Senate, *Concurrent Resolution*, 7 February 1885, and *Report to Accompany Concurrent Resolution*, 7 February 1885. Also *Cong. Rec.*, 48th Congr., 2d sess., Senate, 9 February 1885, 1449–50.

35. U.S. Department of the Navy, Secretary to president pro tempore, 12 February 1885. As already mentioned, in 1850 Congress resolved a controversy by directing the adoption of two initial meridians: the U.S. Naval Observatory's Washington meridian for astronomical and geographical applications, and the Greenwich meridian for all nautical purposes. The statute was not repealed until 1912.

36. U.S. Department of the Navy, Secretary to president pro tempore, 17 February 1885.

37. The advisory committee's first members were the Cambridge astronomer John C. Adams; the astronomer royal W. H. M. Christie; the hydrographer Capt. Sir F. J. O. Evans; the hydrographer of the Royal Navy, Capt. W. J. M. Wharton; Lt. General Richard Strachey; the superintendent of the Nautical Almanac Office, John R. Hind; and the secretary of the Science and Art Department. Though a standing committee, it had no formal name. Sandford Fleming once termed it the "Committee on the Prime Meridian Conference," but this appellation is too narrow. The committee was in existence for at least a dozen years; its importance has remained virtually unrecognized by historians. This committee's actions are discussed in later chapters.

38. More than likely this focus was the result of Christie's early letter urging immediate attention to consideration of universal time and the astronomical day; see Christie to Secretary, 20 April 1885, in U.K. Science and Art Department, *Thirty-third Report* [for 1885]: Appendix A, 27–28. By this date the astronomer royal was aware of both Newcomb's and Foerster's published opposition, to be discussed in the next chapter. For the advisory committee's views, see U.K. Science and Art Department, *Thirty-fourth Report* [for 1886]: Appendix A, 10.

39. An English translation of the imperial ordinance is in U.K. Science and Art Department, *Thirty-fourth Report* [for 1886]: Appendix A, 23. See also Kikuchi, "Time Reform in Japan."

40. The quotation is from President Cleveland's cover letter.

41. De Beaumont, "De la projection" and "Utilité de l'adoption d'un méridien initial." IGC 4, *Compte rendu*, 1:41–42, 163–66, 196–204; BAAS, *Report of the Fifty-ninth Meeting* [for 1889].

42. Sandford Fleming was also a member of the advisory committee during its consideration of the Italian government's invitation; U.K. Science and Art Department, *Thirty-eighth Report* [for 1890]: Appendix A, 27–33. During the International Telegraph Conference in Paris in 1890, de Quarenghi tried to convince the delegates there to include the time defined by the Jerusalem meridian on telegrams; see International Telegraph Conference, *Document de la [1890] Conférence*, 328, 635. His proposal was finessed; however, the mere fact that it was even considered was used to buttress the subsequent arguments presented by the Italian government.

43. My translation; quoted in De Busschere, "L'unification des heures, situation actuelle," 303. De Busschere was chief engineer of the State Railway of Belgium and wrote extensively on the subject of time unification.

44. Without question the most important event at the congress was the proposal by Albrecht Penck, professor of geography at the University of Vienna, for an international map of the world, to be executed at the intermediate scale of 1:1,000,000. However, the actual plan for an integrated series of sheets, with Greenwich as the initial meridian and using the metric system, did not become a reality for eighteen years, during which time parochialism was rife. Significantly, success required the best efforts of a gathering of specialists appointed by their respective governments and empowered to resolve all conflicts. Summaries of

this lengthy process are in Brown, *Story of Maps*, 299–305; Thrower, *Maps and Civilization*, 164–71; and Willis, "International Millionth Map of the World." For the official agreements, see International Map Committee, *Resolutions and Proceedings*, 1–11.

45. Holdich, "Prime Meridian."

Chapter 7

1. The resolution was introduced by Lewis Rutherfurd, whose prefatory remarks are at U.S. President, *Message . . . Relative to International Meridian Conference*, 50. The quotation is from the remarks by delegate Capt. Sir F. J. O. Evans, who until his recent resignation had been the hydrographer of the Royal Navy, and given at ibid., 99.

2. The Admiralty letter, dated 11 October 1805, is reproduced in Norie, *New and Complete Epitome*, 271; NARA, *List of Logbooks*, 111. John C. Adams, director of the Cambridge Observatory, believed the practice was universal, but some merchant ships may have been using this outdated practice in their logs; U.S. President, *Message . . . Relative to International Meridian Conference*, 97.

3. Including the nautical day in Resolution VI appears to reflect a lack of knowledge of shipboard life on the part of a few delegates at Washington; indeed, several of the specialists who met in Rome in 1883 were also unaware of current nautical practice; see U.S. President, *Message . . . Relative to International Meridian Conference*, 99–100, and Hirsch and von Oppolzer, *Unification des longitudes*, 35.

4. The USNO superintendent represented Colombia.

5. Copies of the various orders and communications are in U.S. Congress, Senate, Committee on Foreign Relations, *Letter from the Secretary . . . transmitting communications concerning . . . the Astronomical Day*; major extracts can be found in *Nature*, "The Question of Civil and Astronomical Time." For an appreciation of Newcomb's importance and details of his work, see Dick, *Sky and Ocean Joined*, chapter 8.

6. The quotation is from Dick, ibid., 171. Franklin's autobiography, *Memories of a Rear-Admiral*, sheds no light upon the controversy; indeed, his brief tour of duty at the Naval Observatory is scarcely mentioned.

7. How the superintendent used the advisory board is not clear, for soon after, one member—the senior astronomer Asaph Hall—strongly criticized the proposed change.

8. At the International Geodetic Association's Rome meeting, Christie proposed that Greenwich civil time be adopted by astronomers; the vote against the proposal was five to two. For the announcement of the astronomer royal's subsequent decision, see Ellis, "Prime Meridian Conference," 10. Like his American counterpart, the astronomer royal was not responsible for the Admiralty's *Nautical Almanac*, which was under the superintendence of John R. Hind.

9. Quoted in U.S. Congress, Senate, Committee on Foreign Relations, *Letter from the Secretary . . . transmitting communications concerning . . . the Astronomical Day*, Communication No. 6. Starting in 1885, nonastronomical

observations and spectroscopic and photographic studies of the sun at the Royal Observatory were reported in terms of Greenwich civil time; see Howse, *Greenwich Time and the Longitude*, 150.

10. Secretary of State Frelinghuysen's 5 February 1885 letter to the chairman of the Senate Committee on Foreign Relations was unambiguous in this regard; see the quotation given in chapter 6 regarding President Arthur's opinion.

11. U.S. Congress, Senate, Committee on Foreign Relations, *Letter from the Secretary . . . transmitting communications concerning . . . the Astronomical Day*, Communication No. 7.

12. Ibid., Communication No. 19; the quotation is in ibid., Communication No. 21. Franklin informed Newcomb on 2 January 1885, ibid., Communication No. 20.

13. Newcomb, "On the Proposed Change."

14. In his autobiography, prepared two decades later, Newcomb wrote that in 1886 he and a later Naval Observatory superintendent (George E. Belknap) prepared "a very elaborate report, discussing the subject especially in its relations to navigators, pointing out . . . the danger of placing in the hands of navigators an almanac in terms of civil time with its tables necessarily different from what they were accustomed"—which, given the four decades' use of civil reckoning in Navy ship logs, certainly turns the argument on its head! *Reminiscences*, 228. Dated 31 March 1886, the report is discussed later in this chapter.

15. Newcomb, "On the Proposed Change," 123. In his autobiography Newcomb also noted that Astronomer Royal Sir George Airy, who had retired recently, in 1881, had written, "I hope you will succeed in having its adoption postponed until 1900, and when 1900 comes, I hope you will further succeed in having it again postponed until the year 2000." It is worth mentioning here that, by law, Simon Newcomb had to retire from active government service in 1897.

16. Foerster, "Ueber die von der Conferenz zu Washington"; an English-language summary with comments is in "The Proposed Change in the Astronomical Day," *Observatory* 8 (March 1885): 81–82.

17. G. F. A. Auwers, president of the Astronomische Gesellschaft, and F. Tietjen, editor of the *Berliner Astronomisches Jahrbuch* and the complementary *Nautisches Jahrbuch*.

18. Oppolzer, "On the Proposed Change."

19. Winlock, "The Proposed Change in the Astronomical Day"; Oppolzer, *Canon der Finsternisse*. As the Harvard astronomers D. H. Menzel and O. Gingerich write in their preface, "The book is all the more useful because the author, several decades before its general adoption by astronomers, had the foresight to employ Universal Time based on the Greenwich Meridian."

20. Struve, *Die Beschlüsse der Washingtoner Meridianconferenz*; an English translation is in Fleming, *Universal or Cosmic Time*, suppl. papers, 84–101.

21. The quotations are from Fleming's translation, at ibid., 87, 97.

22. *Science*, "Question of Civil and Astronomical Time"; U.S. Department of the Navy, Secretary to O. C. Marsh, president of the National Academy of Sciences, 22 April 1885.

23. Winlock, "Proposed Change," 524–25. According to Winlock, the as-

tronomers Adams, Christie, and Struve were in favor, while Newcomb, Foerster, and Auwers were opposed.

24. Astronomische Gesellschaft, *Vierteljahrsschrift der Astronomischen Gesellschaft.*

25. The first quotation is from "Proposed Change in the Astronomical Day," *Observatory* 8 (October 1885): 347. The other two quotations are translations of Gautier, "Est-il opportun?" ibid. (December 1885): 423. Over a sixteen-month period the *Observatory*'s editors published more than a dozen articles and remarks on the proposed change in the astronomical day and only a half dozen entries on universal time, a clear indication of the importance of the former to the astronomy community.

26. National Academy of Sciences, "Astronomical Day," 2–4, at 4. The Senate Committee on Foreign Relations had jurisdiction over International Meridian Conference issues.

27. Asaph Hall, "Reports of the National [A]cademy of [S]ciences" and "Proposed Change." As a matter of record, conference delegates included one admiral and one commodore, both American; two generals, one British, one Russian, both well-known in their technical specialties; sixteen diplomats; and sixteen additional technical specialists, including six astronomers.

28. The diplomatic correspondence is in U.K. Science and Art Department, *Thirty-fourth Report* [for 1886]: Appendix A, 10–12. A key piece of correspondence, from the Admiralty and dated 5 January 1886, is in ibid., *Thirty-third Report* [for 1885]: Appendix A, 38–39.

29. The professors of mathematics were Asaph Hall, William Harkness, J. R. Eastman, and Edgar Frisby, at the Naval Observatory; and astronomer Stimson J. Brown, who was then assigned to duty at the Naval Academy. Brown introduced practical astronomy into the discussion. All are quoted in U.K. Science and Art Department, *Thirty-fourth Report* [for 1886]: Appendix A, 15–19.

30. Belknap and Newcomb to Chief of Bureau of Navigation, quoted in ibid., 12–15. I am using this version and also ignoring their reference to the nautical day, for both authors dismiss it as an unimportant consideration.

31. The general opinion was that the French government was opposed to the proposed change. However, that was not the position of the astronomer and geodesist Hervé Faye, longtime president of France's Bureau des Longitudes. At the IGA's 1883 Rome conference this official French delegate voted to make the universal day conform to the civil day, his and Astronomer Royal Christie's votes being the only ones to favor the change. In his subsequent report on the Rome conference's decision to have the universal day begin at noon, Faye repeated his objection to it, reminding his readers that Laplace had used civil time in his seminal *Celestial Mechanics*; see Hirsch and von Oppolzer, *Unification des longitudes*, 34–35; and Faye, "Sur l'heure universelle." That the selection of an initial meridian and the adoption of universal time based on civil reckoning were quite separate issues can be seen in Bureau des Longitudes member Antoine d'Abbadie's "Choix d'un premier méridien," where the astronomer argued for selecting the island of Flores in the Azores as the prime meridian, while at the same time calling astronomical time reckoning that commenced at noon a "complication without advantage."

32. George Wheeler had reached this same conclusion some years before. In his report on the Third International Geographical Conference held in 1881, delegate Wheeler wrote that "[in all nautical and astronomical] tables reference can be made to a single meridian in all the publications"; quoted in Bartky, *Selling the True Time*, 263n15.

33. For Fleming's unsuccessful efforts to have the twenty-four-hour notation adopted widely, see Creet, "Sandford Fleming and Universal Time," 85–86, and Wisely, *American Civil Engineer*, 232–34. Fleming's draft bill is in his papers at the Library and Archives Canada, vol. 102, file 32, and displayed on its Web site. Related material is in Canada, House of Commons, "Return to an Order," 11–14. In the United States, House and Senate bills were introduced in both the 51st Congress, 2d sess., 1891, and the 52nd Congress, 1st sess., 1892. No follow-on committee reports were ever issued.

34. Fleming, "Memorandum"; see the discussion of standard time zones in the next chapter.

35. An astronomer and a leader in the distribution of time signals in Canada, Carpmael served in 1893 as president of the Astronomical Society and Physical Society of Toronto.

36. The circular-letter, Canadian Institute, Joint Committee, "Proposed Change in Reckoning," which includes the first report of the Joint Committee, is in *Transactions of the Canadian Institute*, but without ballot; a copy including ballot can be found in Astronomical and Physical Society of Toronto, Joint Committee, *Transactions of the Astronomical and Physical Society for the Year 1892*. The mailing list was based on a compendium recently published by the Royal Astronomical Society of Brussels. Copies of the Joint Committee's second report, containing the astronomers' data and analyses, and its third one are in the Astronomical and Physical Society of Toronto, *Transactions*, both reprinted in the Royal Society of Canada's 1896 *Proceedings and Transactions*.

37. U.K. Science and Art Department, *Forty-second Report* [for 1894]: Appendix A, 12–20.

38. Ibid., 20–21; a copy of the Admiralty report, dated 30 July 1894, was sent to the Science and Art Department.

39. In U.K. Science and Art Department, *Forty-third Report* [for 1895]: Appendix A, 10, dated 3 August 1894. The quotation is from the 29 September 1894 inquiry letter, in U.S. Department of State, United Kingdom Inquiry Letter.

40. The three-page USNO report and the secretary of the navy's cover letter are in U.S. Department of State, Miscellaneous Letters, RG59, M179, roll 900. The transmittal letter is in U.S. Department of State, Response to the United Kingdom, RG59, M99, roll 51, 24 October 1894. A contrasting, favorable response is Poincaré, "Rapport sur la proposition d'unification," E.1–10. During World War I, when the issue of astronomical time was raised once more, Poincaré's report was translated and printed in *Observatory* 40 (September 1917): 323–27.

41. Fleming, letter to Abbe, 21 May 1895.

42. U.S. Department of State, letter to Cleveland Abbe, 8 January 1895; Abbe, letter to Fleming, 24 May 1895.

43. I have been unable to locate a copy of any almanac bearing this name in British and American libraries and maritime museums; possibly Fleming used "Greenwood's Nautical Almanac" as a kind of shorthand for *Greenwood's Kludonometric Tide Tables for the Lancashire Coast and Bristol Channel* . . . , published during the 1890s.

44. Greenwood, "Unification of Time," *Proceedings and Transactions of the Royal Society of Canada*, 2d ser., 2 (1896), Appendix: A11, A29–41. Greenwood subsequently updated his results, which were summarized: 458 respondents, with 97 percent in favor; ibid., A6.

45. Even as late as January 1897 neither country had responded; U.K., Admiralty, letter from Admiralty secretary to Christie, 19 January 1897.

46. Astronomical and Physical Society of Toronto, Joint Committee, Third Report, quotations at 97 and 99.

47. When one applies the same reasoning to astronomers residing in Germany that Fleming applied to those living in the United States, the inescapable conclusion is that German astronomers opposed the change thirty-one to seven.

48. U.K. Admiralty, letter from secretary to undersecretary of state, 12 December 1895.

49. The appeal is dated 9 April 1896 and is printed in Royal Society of Canada, *Proceedings and Transactions*, 2d ser., 2 (1896), Appendix: A9–12; the quotation is at A9.

50. Dated 9 July 1896, the Admiralty letter is in U.K. Science and Art Department, *Forty-fourth Report* [for 1896]: Appendix A, 7.

51. Fleming was a founding member of the Royal Society of Canada, which has been in existence since 1882. The communication is dated 23 April 1896; in Royal Society of Canada, *Proceedings and Transactions*, 2d ser., 2 (1896), A9–11. The Royal Society of Canada's memorial is dated 27 April 1896, with its general meeting being held on 19 May; at ibid., xxii–xxiii; Appendix: A7–8.

52. Bourinot, "Unification," A3–6, a cover memorandum dated 14 October 1896.

53. Royal Society of Canada, "Unification of Time," *Proceedings and Transactions*, 2d ser., 3 (1897): xxvii–xxx. A copy of the 1 January 1897 memorial of the Royal Colonial Institute is in Astronomical and Physical Society of Toronto, *Transactions for the Year 1897*, 2–3; a related public letter designed to enlist British support is Fleming, "Unification of Time at Sea."

54. Secretary, Royal Society, letter to Honorary Secretary, Royal Society of Canada, 25 May 1897; published in Royal Society of Canada, *Proceedings and Transactions*, 2d ser., 3 (1897): xxix.

55. John A. Patterson, "On the Unification of Time," summary in BAAS, *Report of the Sixty-seventh Meeting* [for 1897], 550–51.

56. Astronomical and Physical Society of Toronto, *Transactions for the Year 1897*; the quotations are at 53, 54, and 69. See also "Report of the [BAAS] Council," in BAAS, *Report of the Sixty-seventh Meeting* [for 1897].

57. According to Creet, "Sandford Fleming and Universal Time," 88, the stagnation of efforts aimed at the adoption of the twenty-four-hour notation on many occasions led Fleming to call for the termination of the ASCE's Committee on Uniform Standard Time. Its last report is dated 13 January 1900.

Chapter 8

1. We have already noted Japan's 1886 decision to adopt a Greenwich-linked civil time, the one direct practical result of the International Meridian Conference.

2. Schram, "Einheitliche Zeit." Like most advocates, Schram was too optimistic regarding the speed at which the adoption would taken place—especially in the United States.

3. Schram, "Actual State," 141.

4. Noted in De Busschere, "L'unification des heures," 301.

5. Schram, "Actual State," 141. Foerster's long-standing opposition was noted in *Ciel et Terre*, "L'heure de Greenwich," 23, and in De Busschere, "L'unification des heures en Europe," 202. Bucholz, *Moltke*, 147, writes that standard time's general introduction was requested by the German Empire's military authorities, which is inconsistent with the accounts cited here.

6. Noted in Pasquier, "L'unification de l'heure," 448, who calls it the "General Assembly of the 'Verein'"; also De Busschere, "L'unification des heures, situation actuelle," 322.

7. As posited by Pasquier, "L'unification de l'heure," 448.

8. Von Moltke's speech before the Reichstag was translated at least twice: into French by Louis De Busschere, and into English by Sandford Fleming. My analysis is based on the text in Moltke, *Gesammelte Schriften*, 7:38–43.

9. Ibid., 38–39; my translation.

10. Ibid., 39; my translation.

11. Ibid., 43; my translation.

12. This and following dates are from two sources. The first is the editorial summary of Pasquier, "L'unification de l'heure." A professor of astronomy at the University of Louvain, Pasquier was an important advocate of uniform time based on Greenwich-linked meridians. It must be noted that in this 1891 article Pasquier holds too positive a view regarding the level of acceptance of Greenwich-based standard times among the "civilized" countries of the world. The second source is De Busschere, engineer-in-chief in the Ministry of the Belgian State Railways, "L'unification des heures en Europe."

13. Pasquier, "L'unification de l'heure," 448–49.

14. De Busschere, "L'unification des heures en Europe," 196–97, 198.

15. The astronomer was Edmund Weiss, founder of the observatory at the University of Vienna. See the discussion, in ibid., 202, indicating that his opposition to enacting a change was based on time consistency in astronomical and meteorological observations.

16. *Nature*, "Universal Time," and *BAA Journal*, "Universal Time."

17. Bouquet de la Grye, "Note sur l'adoption."

18. The introduction of the bill is given in Pasquier, "La Belgique et l'heure de Greenwich," 503–5. I suspect the arguments presented in its favor were based on the text of the report of the Bureau des Longitudes–sponsored commission.

19. Bureau des Longitudes, "Unification de l'heure en France," 779.

20. Forel, "L'unification de l'heure." Forel resided near the French border and thus was well aware of the various differences.

21. Ibid., 327–28.

22. De Nordling, "L'unification des heures," 115–16; this work was the result of a talk he had presented to the Geographical Society at its 21 February 1890 session. An important European railway engineer, de Nordling (also known as Wilhelm von Nördling) was writing about railway time and time uniformity in France as early as 1888; his criticisms and his detailed history of events were widely known.

23. De Nordling, "L'unification des heures" and "Le projet de loi."

24. In De Busschere, "L'unification des heures, situation actuelle," 314. In this lengthy article De Busschere quotes Faye's remarks in detail.

25. Ibid., 319–20.

26. "L'heure légale, en France et en Algérie, est l'heure, temps moyen, de Paris."

27. See, for example, *Nature*, "France and International Time."

28. De Nordling, the civil engineer whose dissent to Paris time had been dismissed out of hand by Faye, the president of the Bureau des Longitudes, during the 1891 debates in the senate, extracted some justified revenge in 1893 when he reminded his French Geographical Society audience of Faye's "not even in a hundred years" remark as he listed the European countries that had adopted Greenwich-linked timekeeping; see "Les derniers progrès de l'unification de l'heure."

29. Two additional sections of the draft bill concerned details of the law's application. The classic source for events through 1911, including citations to the official legislative journals, is Bigourdan, "Le jour et ses divisions." I am drawing upon this material here.

30. During this period the most important advocate for modifications was Charles Lallemand of the Bureau of Mines, who in two articles—"L'unification internationale" and "L'heure légale en France"—systematically countered the objections being raised by specialists arguing against any and all changes. (A few of Lallemand's 1897 arguments were translated from an offprint in Galison, *Einstein's Clocks*, 160–61.) Lallemand's views were strongly opposed by the Bureau des Longitudes' Bouquet de la Grye; see his "L'heure nationale."

31. "L'heure légale en France et en Algérie est l'heure, temps moyen de Paris, retardée de 9 minutes 21 secondes"; my translation.

32. Translated from the quoted text in Bigourdan, "Le jour et ses divisions," B.63.

33. See, e.g., the summary in *La Nature*, "L'heure légale française," which highlights the opposition voiced by several members of the Académie des Sciences during its 28 March 1898 session.

34. It must be noted here that by 1910 the European situation had changed.

Spain adopted Greenwich time on 1 January 1901, the last of all of France's direct neighbors to drop their non-Greenwich-based local and national times. On the other hand, in 1908 the Netherlands, which had adopted Greenwich time for its railways and telegraphs in 1892, chose Amsterdam time as its legal time (about nineteen minutes and thirty-two seconds ahead of Greenwich). I am indebted to Robert H. van Gent of Utrecht for the details of his country's public time activities.

35. Lallemand had been one of the French government's representatives to the First International Conference on the International Map, which met in London in November 1909 under the aegis of the British government. For a brief discussion of this important unification activity, see chapter 6, n44.

36. Bigourdan, "Le jour et ses divisions," B.65.

37. The issue of daylight saving time in the United Kingdom is discussed in detail in chapter 10. Cited in Cathenod, "L'heure fuselaire," 212, col. 2.

38. From the summary given in Bigourdan, "Le jour et ses divisions," B.65–66.

39. Ibid., B.67–68.

40. "Time, Actual and Legal," *Times* (London), 14 May 1880, 14; cited in Howse, *Greenwich Time and the Longitude*, 114–15. For the railways and Greenwich (London) time, see Bartky, *Selling the True Time*, 218n4.

41. U.K., *Public General Statutes*.

42. The base meridian in Ireland was that of the Dunsink Observatory; see Wayman, *Dunsink Observatory*, 138. The use of Dublin time in Ireland ended on 1 October 1916.

43. *Nature*, "Time Standards of Europe"; "France and International Time," 3 August 1893; and "France and International Time," 19 May 1898.

44. Christie to secretary, Science and Art Department, 20 April 1885, a letter sent at the time of the advisory committee's formation; in U.K. Science and Art Department, *Thirty-third Report* [for 1885].

45. Christie, "Universal Time."

46. Ibid., 389–90, 391, 393.

47. U.K. Science and Art Department, *Thirty-fourth Report* [for 1886]: Appendix A, 25–26, 27, 28. Note the curt rejection of these various proposals by the Public Works Department of the Government of India.

48. Fleming, "Memorandum on the Movement." Dated 20 November 1889, these materials were printed in *Transactions of the Canadian Institute*, and subsequently in U.K. Science and Art Department, *Thirty-eighth Report* [for 1890]: Appendix A, 16–24. They were also reprinted in the Canadian government's 1891 *Sessional Papers*. This memorandum is apparently the first work in which Fleming supported the adoption of time zones.

49. U.K. Science and Art Department, *Thirty-eighth Report* [for 1890]: Appendix A, 25, with date of 26 July 1890. A copy of the Colonial Office's dispatch, dated 21 November 1890, is in Canada, House of Commons, "Return to an Order," 2.

50. Apparently the exertions of David Gill, H.M. Astronomer at the Cape, led to the result; see Gill, letter to secretary of the Admiralty. The Cape Colony

and the Orange Free State actions are reported in *BAA Journal*, "Unification of Time at the Cape"; also *BAA Journal*, "Uniform Time for South Africa."

51. *Nature*, "Standard Time in Australia." For a map showing the continent's zones, see *BAA Journal*, "Change of Time in Australia." For India, see *BAA Journal*, "A Standard Time for India." New Zealand had adopted a Greenwich-linked time even earlier (1868); see Pawson, "Local Times and Standard Time in New Zealand."

52. Zone time was still not an unqualified success. The Netherlands—see n34 above—did not revert to a Greenwich-linked time until World War II. Chile renounced its Greenwich-linked time on 1 August 1916 but changed back sometime later. As of this writing (2006), the transformation to a unified, worldwide system is still not complete; see the epilogue for a few remarks on uniformity.

Chapter 9

1. Aitken, *Syntony and Spark*, 232, 239, 290n48, 308. Aitken, 110–15 and nn, suggests that a popular article written in 1892 by the prominent British chemist and physicist William Crookes may have stimulated people to think about the applications of wireless technology, including many fanciful ones (i.e., unprofitable in the economic sense).

2. Grubb, "Proposal"; Bigourdan, "Sur la distribution de l'heure"; Normand, "Sur le réglage des montres"; Albrecht, "Über die Verwendbarkeit der draht-losen Telegraphie."

3. USHO, "U.S. Naval Wireless Telegraph Services." Regarding the inaugural date, *Annual Reports of the Navy Department for the Year 1904*, 17–18, 382, and *Annual Reports . . . for the Year 1905*, 320, list six shore stations transmitting wireless signals linked to the Naval Observatory's time service. And in an official document at the 17 October 1912 session of the International Conference on Time submitted by the American delegate, Captain H. H. Hough, U.S. Navy, it was stated that the wireless time service commenced in January 1905 and continued without interruption; see Bureau des Longitudes, "Conférence Internationale de l'Heure," D.84.

4. Hutchinson, "Wireless Time Signals"; the author is incorrect in his assertion that this 1907 time service was the first in the world.

5. Bigourdan, "Le jour et ses divisions," B.77–80.

6. It should be noted that the time signal to be transmitted was to determine the error and rate of the ship's chronometer, the onboard source of time at the initial meridian.

7. Bouquet de la Grye, "Détermination de l'heure"; Guyou, "Détermination des longitudes en mer." Also Bigourdan, "Sur l'application de la télégraphie sans fil."

8. The quotation is my translation of the French text in Bigourdan, "Le jour et ses divisions," B.76, where it is italicized.

9. Galison, *Einstein's Clocks*, 276–77, states that in the winter of 1908 an Interministerial Technical Committee for Wireless Telegraphy was established.

Chaired by Henri Poincaré, the committee received various technical proposals and monitored progress.

   10. Ferrié, [Report on the Transmission of Time Signals]. Galison, *Einstein's Clocks*, 276, indicates this report was presented to the interagency committee on 8 March 1909.

   11. Bigourdan, "Le jour et ses divisions," B.78–79.

   12. Poincaré, "Sur les signaux horaire"; see also n7 above.

   13. As already noted, Bigourdan, "Le jour et ses divisions," B.65, gives the date of the government's acceptance of the bill, "sans réserve," as 20 July 1910.

   14. Cathenod, "L'heure fuselaire."

   15. Bureau des Longitudes, "Congrès international des éphémérides astronomiques," A.3–4. Baillaud was president of the Permanent International Committee of the Photographic Chart of the Sky, the famous *Carte de Ciel*, a long-term project instituted by the Paris Observatory in 1887; the correspondence was sent under its aegis. At the International Conference on Fundamental Stars held in Paris in May 1896, almanac office directors had agreed on standards for stellar catalogs and particular constants to be used in the construction of tables.

   16. See the remarks of G. B. Boccardi, director of the Turin Observatory; Fritz Cohn, director of the *Berliner Astronomisches Jahrbuch* and the Royal Institute for Astronomical Calculations; and C. D. Perrine, director of the National Observatory at Cordoba, Argentina; in Bureau des Longitudes, "Congrès international des éphémérides astronomiques," A.8–10, 11.

   17. Ibid., A.21–23, especially Poincaré's comments, representing the views of the Bureau des Longitudes.

   18. The resolutions are at ibid., A.41–44; the quotation is at A.19. In the United States, sharing both the data and the workload as well as dropping the Washington meridian required approval from Congress, which was forthcoming; see U.S. Department of the Navy, *Annual Reports . . . for the Fiscal Year 1912.*

   19. The realization came only after time signals from multiple observatories were compared at a central location on a regular basis; see Bartky, "Gauging Time Accurately," in *Selling the True Time*, especially figure 9.1.

   20. Ferrié, "Sur quelques nouvelles applications," 181–82.

   21. A comparison program began at the Royal Observatory, Greenwich, on 5 July 1912, but only via a receiving station. It is noteworthy that no mention was made of radio time signals linked to the Royal Observatory until well after World War I had ended. Howse, *Greenwich Time and the Longitude*, 155, suggests that in times of peace time balls were considered adequate, and in times of war "radio time signals would be stopped." His reasoning seems forced, particularly when one considers the enormous advantage to battle fleets of being able to regulate their chronometers while at sea; indeed, British warships received radio messages and instructions all the time. More likely, time signals *were* part of the Royal Navy's message traffic, with the Greenwich time signals sent to its transmitting station via telegraph land lines. In the case of the U.S. Navy, no mention of observatory-telegraphed time signals is found in any of its annual reports until 1912, even though several Navy radio stations had been transmitting U.S. Naval Observatory time for seven years.

22. Said to be in a report by Major Ferrié, but noted without citation by Baillaud and Chandon in their "Rapport relatif aux signals horaires," 135. These authors also wrote that Ferrié proposed establishing a Central Bureau of Time, which appears not to have been the case. Unfortunately, this paragraph has been misread by a host of subsequent authors, among them Lambert, "Le Bureau international de l'heure," 2; and Audoin and Guinot, *Measurement of Time*, 43–44. All wrote that the discovery of discrepancies led to the creation of the International Bureau of Time (BIH). As shown in this chapter, others well before Ferrié had been urging the creation of an international time bureau; thus, Ferrié's remark, purportedly made in late 1911, cannot be viewed as the triggering event.

23. A major role in both identifying and resolving the issue was undertaken by the astronomer royal for Scotland, Robert A. Sampson, via the auspices of the International Astronomical Union; see Dick, *Sky and Ocean Joined*, 456.

24. Guinot, "Le Bureau international de l'heure," 32; my translation.

25. As quoted in the *Observatory*, [telegram to *Titanic*].

26. Bureau des Longitudes, "Conference internationale de l'Heure."

27. Held in June, the recommendations were approved in July by the countries that were signatory to the International Radiotelegraphic Convention.

28. Hall, Jr., Manuscript Trip Report.

29. In Bureau des Longitudes, "Conference internationale de l'Heure," D.82, 261–68.

30. Since the focus is on adoption of the Greenwich meridian for time, which took place prior to World War I, the postwar history of international time services is not discussed here. See instead Paris Observatory, *Le Bureau international de l'heure*.

31. The decision's date was 19 August 1913, as given in Bureau des Longitudes, "Longitude," and in Renaud, "La carte marine internationale," 38.

32. Spain adopted the Greenwich meridian for its navy on 3 April 1907, and Portugal had adopted Greenwich by 1913. I am indebted to Carlos Nélson Lopes da Costa, Technical Director, Hydrographic Institute, Lisbon, for examining two large-scale 1913 hydrographic charts in the institute's archives in order to determine Portugal's date of change.

33. Lallemand, "Sur un projet de Carte internationale"; "Cartes et repères aéronautiques." For a summary of early uniformity efforts, see Ehrenberg, "'Up in the Air.'"

34. Chasseaud, *Artillery's Astrologers*, chapters 1–3.

35. See the next chapter, "Advancing Sunset, Saving Daylight."

36. The time being kept by marine chronometers and deck watches for purposes of navigation was of course to be excluded by Renaud in this and in all subsequent discussions.

37. Renaud, letter to Parry, 29 January 1917; Parry to Renaud, 19 March 1917. I am most indebted to Jonathan Betts, Senior Curator of Horology, National Maritime Museum, Royal Observatory Greenwich, for drawing my attention to this important material and for making his notes available to me.

38. Renaud, "L'heure en mer," 15 January 1917, at B.31. A translation of Renaud's analysis, part of the "Memorandum by Monsieur Renaud," is in UKHO Archives, "Time Keeping at Sea."

39. Renaud, "L'heure en mer," B.5–8; see also the translation in UKHO Archives, "Time Keeping at Sea."

40. Ibid., B.15–22; the quotation is from the translation, 18.

41. Ibid., B.30–31; the quotation in the text is from "Memorandum by Monsieur Renaud," 26.

42. Renaud, "L'heure à bord des navires." Translations of the "Report of the Bureau des Longitudes," 14 February 1917, and "Ministerial Circular Altering the Method of Time-Keeping at Sea," 22 March 1917, are in "Memorandum by Monsieur Renaud," Addendum I. The Bureau des Longitudes report is quoted in Lallemand, "L'heure à bord des navires."

43. France, *Journal officiel de la République Française*, 29 and 30 March 1917, briefly summarized in Royal Geographical Society, "Standard Time at Sea"; France, Service Hydrographique de la Marine, "Limites des fuseaux horaires," and Carte no. 7, "Planisphère."

44. Admiralty (represented by the astronomer royal), Board of Trade, General Post Office, Home Office, Meteorological Office, Ordnance Survey and War Office; and the Royal Society, Royal Astronomical Society, Royal Geographical Society, and Eastern Telegraph Company.

45. Parry, "Opening Statement by the Chairman."

46. UKHO Archives, "Conference on Time-Keeping at Sea." It is possible that the initial report on the conference to the lords of the Admiralty was not available until late in 1917.

47. Ibid., 4.

48. On 22 June 1917; see Renaud, "L'heure en mer," Annex IV, B.39.

49. Ibid., B.19; see also "Memorandum by Monsieur Renaud," 15.

50. U.K. Board of Trade, Marine Division, "Memorandum on Time Keeping at Sea."

51. The quotation is from UKHO Archives, "Conference on Time-keeping at Sea," 3. Sadler, "Mean Solar Time," 295, notes objections by the representatives to the Admiralty committee from the Ordnance Survey and the Royal Geographical Society.

As noted in n42 above, at the Bureau des Longitudes meeting on 14 March 1917 responding to a proposal by Charles Lallemand and Joseph Renaud, the bureau decided to ask the minister of marine for his views regarding the substitution of civil time for astronomical time in its *Extrait de la Connaissance des temps* (Abridged Nautical Almanac). A report was prepared giving the reasons for doing so; it received the approval of the minister on 2 April 1917. As a result, the Bureau des Longitudes announced that the shift to civil time would take place in the 1920 issue of the abridged almanac and informed the Admiralty of the decision. See Renaud, "L'heure en mer," translated as "Memorandum by Monsieur Renaud," addendum III.

52. Lallemand, "À propos de l'extension"; Dyson and Turner, "Commencement."

53. These and subsequent dates are taken from Sadler, "Mean Solar Time," who emphasizes the actions of the RAS. It must be noted that the article is flawed in several respects. Sadler seems unaware of actions that took place prior to the Conference on Time-keeping at Sea. For example, Renaud is identified only as a senior representative of the French Navy; the fact that the Royal Navy had been using Greenwich Time and not apparent time in the Home Water zone seems to have escaped him; and he quotes the recommendation regarding the use of Greenwich Time for wireless and other messages incorrectly. Matters also become somewhat confused when he misquotes the conference's recommendation regarding a change to civil time, writing "the general substitution of the civil for the nautical day" instead of "for the astronomical day." Also, in several places Sadler refers to the nautical day, suggesting that he may not have known that by Admiralty directive the nautical day had been abandoned in 1805 in favor of the civil day.

54. Ibid., 294.

55. Royal Geographical Society, "Standard Time at Sea"; for the Board of Trade's memorandum and accompanying chart, see n50 above; Renaud, "Le Planisphère," with a chart that includes a zone scale similar to the Admiralty's choice for determining universal time; Lallemand and Renaud, "Substitution du temps civil"; Crommelin, "Time at Sea" and "New System."

56. Astronomer Royal Christie's actions are mentioned in a brief but cogent summary of events; Dreyer and Turner, *History of the Royal Astronomical Society*, 225–26.

57. RAS, "Proposed Change." Sadler terms the Admiralty letter, which is dated 8 June 1918, "not very helpful"; see Sadler, "Mean Solar Time," 294.

58. Sadler, ibid., 295, gives some partial statistics, including an analysis by the secretary of the RAS. In my judgment these data and the conclusions based on them are of no real importance. In addition, Sadler's view of the importance of the RAS's activities compared with those of the Admiralty is at variance with mine.

59. As quoted in Sadler, ibid.

60. Sadler, ibid.; RAS, "Commencement of the Astronomical Day"; the letter is dated 19 February 1919. U.K. Admiralty, "Method of Expressing Astronomical Time." The issue of nomenclature regarding the use of Greenwich Mean Time also came to the fore; for an appreciation of the issue among astronomers, see Sadler, "Mean Solar Time," 295, 296–98, 300–301, 303–4.

61. U.K. Admiralty, Admiralty Order no. 1291. See U.K. Hydrographic Department, "World Time Zone Chart," published in anticipation of their lordships' decision. The subsequent directive issued was U.K. Hydrographic Department, "System of Time-keeping at Sea"; signed by F. C. Learmonth, Hydrographer of the Navy, it referenced Admiralty Chart no. D6. This series of charts was the predecessor to the long-running Hydrographic Office series no. 5006, with revised editions still being issued; see U.K. Hydrographic Department, "Time Zone Chart of the World."

62. U.S. Department of the Navy, "Use of Standard Time Zones at Sea"; USHO, "Time Zones." The U.S. Hydrographic Office chart series was H.O.

5192, first published May 1920 and now published periodically by the National Geospatial-Intelligence Agency as "Standard Time Zone Chart of the World."

63. It must be noted that Renaud's efforts in the direction of uniform time were in the context of uniformity in hydrography: one of his last major essays was "La carte marine internationale." His death in 1921 at the age of sixty-six was seen as an enormous loss to the field. To him, more than to any other, belongs the credit for the formation of the International Hydrographic Bureau, now the International Hydrographic Organization, Monaco; see Parry, "History of the Inception"; and Moitoret, *International Hydrographic Bureau.*

Chapter 10

1. Franklin, "A Subscriber to [the Journal's] Authors." Originally in French and untitled, Franklin's letter subsequently received a variety of labels, including "An Economical Project," "Turkey versus Eagle, McCauley Is My Beagle," and "Daylight Saving." The translation used here is "An Economical Project."

2. In 1785, under Louis XVI, the livre tournois contained 4.5 grams of fine gold (99.99%). In April 2006 gold futures were over $600 USD per troy ounce, the equivalent of $19.29 per gram.

3. In October 1895, before the Wellington Philosophical Society, the postal clerk, entomologist, and amateur astronomer George V. Hudson (1867–1946) proposed that New Zealand's clocks be advanced two hours during October through February to gain more light in the evening. (The hours-of-sunlight curve in the Southern Hemisphere is opposite that in the Northern Hemisphere.) Hudson's paper on the subject was distributed in Christchurch, and in 1898 he raised the issue a second time. A bill to advance all clocks one hour was introduced in the country's House of Representatives in 1909 and in every subsequent session for the next twenty years. Nevertheless, there appears to be no link between Hudson's proposal and Willett's more famous efforts. Hudson, "On Seasonal Time." See also *Nature* 78 (4 June 1908): 108, where Hudson's pamphlet was bought to the readers' attention.

4. Biggs, "Willett," *Dictionary of National Biography*; Churchill, "A Silent Toast," first published in *Pictorial Weekly* (28 April 1934); de Carle, *British Time*, chapter 9; Waymark, *History of Petts Wood*, chapter 3. Willett was elected to membership in the Royal Astronomical Society on 10 May 1895.

5. To gain confidence in the accounts of Willett's biographers, for the builder apparently never wrote down how he came to his idea, we examined a large-scale (six inches to the mile) Ordnance Survey map. From 1894 until his death in 1915, Willett lived at The Cedars, located on Camden Park Road at its intersection with a number of streets, including Watts Lane, the most direct route to St. Paul's Cray Common and Petts Wood. From Willett's residence to the wood is under a mile, with a church, a manor house, and fewer than a dozen homes along the way. (A slightly less direct route would add a few more houses.)

6. Willett, *Waste of Daylight*; an edited version, with additions, appeared in the *Horological Journal*. The quotations here are from the original text, as given by de Carle in *British Time*.

7. In the Northern Hemisphere the sun sets at 6:00 P.M. or later on every day between 21 March and 21 September. (In the Southern Hemisphere, the span is from 21 September to 21 March.) Willett selected the first Sunday in April and the fourth Sunday in September for his advanced-time period in recognition of these astronomical limits.

8. At the latitude of London (51°30'N), Willett's proposal would lead to more than fifty sunrises later than 6:00 A.M., most of them occurring during August and September.

9. Willett, *Waste of Daylight*, [April] 1908. In this version Willett includes a skeleton table showing both sunsets and sunrises and how they would change as a result of his proposal.

10. Willett, *Waste of Daylight: Extracts from Letters and Newspapers*.

11. During his 1908 testimony, Willett listed 139 periodicals in favor, 10 adverse, and 87 critical ones. Despite this evidence, de Carle, *British Time*, 157–58, writes: "At first Willett's scheme was met with ridicule[;] the Press—with few exceptions—was loud in its protestations."

12. Willett, *Waste of Daylight*, [April] 1908, 12–15.

13. "Daylight Saving Bill," *Times* (London), 27 March 1908, 10; "Political Notes," ibid., 4 March 1909, 12.

14. Ibid., "Daylight Saving Bill," 8 May 1908, 23.

15. For an example, see Willett's letter to the editor in the *Horological Journal*. For the list of articles, see United Kingdom, Parliament, House of Commons, *Report, and Special Report, from the Select Committee on the [1908] Daylight Saving Bill*, Appendix B; see also the summary in n11.

16. United Kingdom, Parliament, House of Commons, *Report on the [1908] Daylight Saving Bill*. Included are a 161-page "Minutes of Evidence" and five appendices.

17. "Daylight Saving Bill," *Times* (London), 3 June 1908, 17.

18. The quotations are from "The Daylight Saving Bill," *Nature*.

19. When Summer Time was adopted in 1916, a notice was printed in the newspapers requesting that the kingdom's volunteer meteorological observers continue to record their data on Greenwich time. However, rainfall totals were recorded at 9:00 A.M. (GMT), thus 10:00 A.M. Summer Time. This was quite late for the many volunteers who had to be at work by that time. In this case the observer was directed to make the observation at 9:00 A.M. Summer Time and note the change in his daily record; "Rainfall and Summer Time," *Times* (London), 13 May 1916, 7. The quotation is from "Economy in Sunshine," *Times* (London), 6 May 1916, 9.

20. Regarding this opposition, see Wright, letter, 14 September 1907, and additional letters reprinted in the *Horological Journal* 50 (1908): 22–26. Wright, the vice president of the British Watch and Clockmaker's Guild, testified before the Select Committee; his remarks are summarized in *Horological Journal* 50 (1908): 184.

21. United Kingdom, Parliament, House of Commons, *Report, and Special Report, from the Select Committee on the [1908] Daylight Saving Bill*, iii, iv.

22. "The Daylight Saving Bill," *Times* (London), 3 July 1908, 13. Between

4 and 28 July the newspaper printed about three dozen letters to the editor on this subject.

23. A useful summary is Pollak, "'Efficiency, Preparedness and Conservation.'" Another summary, this one rather contemptuous of the legislation's opponents, is Turner, "Taking Time by the Forelock," in *Roads to Ruin*.

24. Expanded English editions of *Waste of Daylight* were printed in 1908, 1911, 1912, and 1914; only in the last two did Willett drop the four 20-minute shifts in favor of a single shift of 1 hour.

25. "Would Add an Hour to Our Summer Day," *New York Times*, 7 February 1909, 3:3. Willett may have been successful with the U.S. Congress, for bills were introduced in May; see the following chapter. In addition, Willett's plan was highlighted in simultaneous articles: "'Saving Daylight': A New Issue in England," *New York Times*, 28 March 1909, 5:5; and "Turn the Clock Back: An English Plan to Economize on Daylight," *Washington Post*, 28 March 1909, 14.

26. From a summary of Churchill's remarks: "The Daylight Saving Bill," *Times* (London), 6 March 1909, 11.

27. United Kingdom, Parliament, House of Commons, *Report and Special Report from the Select Committee on the [1909] Daylight Saving Bill*. The report includes 215 pages of "Minutes of Evidence" and 23 additional pages made up of ten appendices.

28. "Mr. Churchill on Daylight Saving," *Times* (London), 4 May 1911, 7. Churchill's speech was interrupted several times by male and female suffragists protesting the government's position on rights for women.

29. "Daylight Saving," *Times* (London), 5 May 1911, 9.

30. By Willett's count, nineteen separate editions were published, a number that may include non-English versions.

31. The French edition is *Le gaspillage de la lumière du jour*. I have been unable to locate a copy of the German edition, which was apparently first published in 1909 or 1910.

32. Willett, Letter.

33. Hassinger, "Für oder gegen die Sommerzeit?" In an excellent summary, the French hydrographer Joseph Renaud wrote that the German estimate was 200 million (francs), and 100 million (francs) in the case of Austria; see Renaud, "L'avance de l'heure légale."

34. "Notes," *Nature* 97 (6 April 1916): 126–27; "The Daylight Saving Scheme," ibid. (27 April 1916): 183–84; "M[r]. Ch. Lallemand on Daylight Saving in France," ibid. (4 May 1916): 209; "The Daylight Saving Scheme," ibid. (11 May 1916): 221; "Daylight and Darkness," ibid. (11 May 1916): 222–24; "Notes," ibid. (1 June 1916): 283; and "The Paris Correspondent of the *Times*," ibid. (8 June 1916): 308.

35. See n48 below.

36. Pollak, "'Efficiency, Preparedness and Conservation,'" 8–9. Pollak writes that the subject was introduced in Germany and Austria in February 1916, indicating a similar legislative gestation period.

37. "Discussion at the Chambers of Commerce Conference," *Times* (London), 11 June 1914, 21. Discussed in Renaud, "L'avance de l'heure légale," B.5–12.

38. The summaries of France's legislative process and the views of the various groups given here are based on Renaud, "L'avance de l'heure légale," B.12–64.

39. Lallemand, "Sur un projet de modification"; summarized in "M[r]. Ch. Lallemand on Daylight Saving in France," *Nature* 97 (4 May 1916): 209.

40. Lallemand, "Sur un projet de modification," 540.

41. The work of this secret committee, leading to the Académie's official position, was revealed sometime later, when the minister of public instruction, fine arts, and inventions of interest to the national defense cited it during subsequent legislative deliberations. For details, see *Le Génie Civil*, "Le projet de modification"; *La Nature*, "Modification de l'heure légale"; and Renaud, "L'avance de l'heure légale," B.28.

42. Renaud, "L'avance de l'heure légale," B.28–32. It should be noted that the Bureau des Longitudes was part of the Ministry of Public Instruction; thus, its position ran counter to that taken by the minister. Moreover, the bureau's quoted position, as given by Renaud, was something of an exaggeration and misdirection. The actual issue was that countries that in 1913 had attended a diplomatic *convention internationale* on the subject of time agreed to create an International Association on Time. Ratification of the convention by the signatories was a necessary next step; however, the war made that course of action impossible.

43. Renaud, "L'avance de l'heure légale," B.35.

44. *Le Génie Civil*, "Le projet de modification"; P. D., "Les économies de combustible"; A. P., "L'économie d'énergie électrique." During 1917 and 1918, and after the Armistice, France observed advanced time for periods of six and seven months.

45. United Kingdom, Home Department, Summer Time Committee, *Report of the Committee*. An edited version of this report is in U.S. Senate, Committee on Interstate Commerce, *Daylight Saving Time and Standard Time for the United States: Hearings on S. 1854*, 49–66.

46. "Economies in Light and Fuel," *Report of the [Summer Time] Committee*, 10.

47. "Agriculture," ibid., 14.

48. The separate focus on Ireland, under a new time since 1 October 1916, may be related more to ongoing political events—the Easter 1916 uprising in Dublin and its aftermath—than to Summer Time concerns, an idea raised by Lane Jennings. Except for one year, 1917, this recommendation was ignored.

49. "Conclusions," *Report of the [Summer Time] Committee*, 18.

50. Summarized in United Kingdom, Home Office, Scottish Home and Health Department, *Review of British Standard Time*. "British Standard Time" is what Americans call "Year-Round Daylight Saving Time."

51. Waymark, *History of Petts Wood*, chapter 4. The words "Erected 1927" were incised on the tablet's north face and "Summer Time Act 1925" on the south face.

Chapter 11

1. Bartky, *Selling the True Time*, 144–46.

2. From *Popular Astronomy*, "Detroit's Different Kinds of Time," quoting an article in the 15 December 1900 *Literary Digest*. The 1905 date is in Howse, *Greenwich Time and the Longitude*, 125 (quoting a 1938 newspaper).

3. Early on, the Detroit More Daylight Club wrote to William Willett. In his 25 May 1909 testimony before the British Parliament's Select Committee, he mentioned that the organization had ten thousand members, all of them supporting the "More Daylight Scheme."

4. Described in O'Malley, *Keeping Watch*, 265, with citations. Also Prerau, *Seize the Daylight*, 74–78.

5. Ordinance 4326, passed 7 February 1890. I am indebted to David P. Whelan, Director, Cincinnati Law Library Association, for this and the additional information in n9 below.

6. H.R. 10028, 61st Congr., 1st sess., "A Bill to Establish a Variation of the Measurement of Time," and S. 2485, 61 Congr., 1st sess., "A Bill Establishing a Universal Standard of Time," were introduced on 24 May and 27 May, respectively. The bills had identical texts; if passed, the legislation would have been termed "The Daylight Savings [*sic*] Act, Nineteen Hundred and Nine." See also *Congr. Rec.*, 61st Congr., 1st sess., 24 May 1909, 44:2347; and ibid., 27 May 1909, 44:2427. The activities of the association during this period were reported in the *New York Times*, the *Washington Post*, the *Washington Evening Star*, and the *Times* (London).

7. "Adopts Longer Afternoons," *New York Times*, 29 June 1909, 1; "Clocks Will Be Turned Back [*sic*]," *Cincinnati Enquirer*, 29 June 1909, 14. Also "Daylight Saving in Cincinnati," *Times* (London), 3 July 1909, 5, an account leading to Willett's mentioning the association's activities in Cincinnati before Parliament's Select Committee during his 6 July 1909 testimony.

8. "Adopts Longer Afternoons," *New York Times*, 29 June 1909, 1; "Why Not Trust Instinct?" ibid., 2 July 1909, 6; and "Clocks Are All Right as They Are," ibid., 3 July 1909, 1. The newspaper's editors felt that for such a change to be effective, everyone had to switch at the same time, not piecemeal. The railroader William Allen penned a letter to the editor in which he voiced great opposition to the shift; see "'Greater Daylight' Plan," *New York Times*, 2 July 1909, 6. His date regarding Cincinnati's adoption of Central time is incorrect; see n5 above.

9. Repealed by Ordinance 1825, signed by the mayor on 14 April 1910; see also "City Ordinances," *Cincinnati Enquirer*, 20 April 1910, 11.

10. "To Boost Exchange Clock," *New York Times*, 4 May 1916, 16. At this moment Great Britain had just started to consider advancing its clocks, mindful of the London Stock Exchange's long-held objection to losing the hour's overlap with New York's financial markets. After the kingdom's switch to Summer Time later in the month, the overlap between the two markets vanished. How significant this loss may have been is impossible to determine; one New York spokesman said early on that it "wouldn't make a great deal of difference";

"Kaiser's Trick Now Gives Us a Problem," *New York Times*, 8 May 1916, 8. Of course, once advocacy in favor of advancing clocks in the United States was in full swing, this particular loss was often cited.

11. "Many Favor Clock Change," *New York Times*, 30 May 1916, 5. The article noted a bill "recently introduced in Congress" by Representative William Borland of Missouri. However, no such bill had been introduced by anyone at this stage; Borland himself did not introduce a daylight-saving-time bill until January 1917.

12. This twenty-two-month period was described in O'Malley, *Keeping Watch*, 267–78. Several of the dates he cites are incorrect, including ones associated with events in Congress. In addition, Robert Garland of Pittsburgh, who spearheaded the effort in the Chamber of Commerce of the United States, is not included in O'Malley's discussion. Further, he significantly overstates A. Lincoln Filene's role in the lobbying effort and apparently confuses him with his brother Edward, for he calls the former a leader in the Chamber of Commerce of the United States. Moreover, O'Malley alleges that increased profits for the brothers' Boston store was the underlying reason for Lincoln Filene's participation in the daylight-saving-time movement, while admitting that he lacks any direct evidence for his allegation; ibid., 292.

13. 64th Congr., 1st sess., S. J. Res. 135, introduced 25 May 1916; and H. J. Res. 229, introduced 26 May 1916.

14. The Chicago Association of Commerce established a Special Committee on Change of Time, which presented its report in December 1914; see Chicago Association of Commerce, "A Longer Sunlight Day." The report was accepted and presented at the annual meeting of the Chamber of Commerce of the United States in February 1915; however, no action was taken there.

15. Similar to Detroit's 1915 adoption of Eastern Standard Time, Cleveland's actions are noted in O'Malley, *Keeping Watch*, 264, 266–67, and in Prerau, *Seize the Daylight*, 76, 77.

16. As quoted during the 26 May 1916 Senate hearing on S. J. Res. 135; see n17.

17. The record of the May 1916 hearing, *Standardization of Time*, 64th Congr., 1st sess., which includes Chicago Association of Commerce, "Report of Special Committee on Proposed Change of Time," is printed in U.S. Senate, Committee on Interstate Commerce, *Daylight Saving Time and Standard Time for the United States: Hearings on S. 1854*.

18. "Are We the Slaves of Clocks?" *New York Times*, 8 May 1916, 10.

19. This early history is from Garland, *Ten Years of Daylight Saving*. Committee members included Eugene U. Kimbark, the 1914 chairman of the Chicago Association of Commerce's Special Committee on Change of Time, already discussed; and A. Lincoln Filene, chairman of the Boston Chamber of Commerce's Special Committee on [a] Daylight Saving Plan. Marcus M. Marks attended the meeting in Pittsburgh as president of the New York Daylight Saving Committee.

20. At some stage in the process, the Harvard University astronomy professor Robert W. Willson was added to the committee. He calculated sunrises

and sunsets for cities in the Eastern time zone for the report. Later his calculations for cities throughout the United States were entered into the hearing record for the daylight-saving bill that was eventually enacted (S. 1854); see n28 below.

21. Accounts of the special committee's efforts appeared in *Current Affairs* almost weekly from 14 June 1916 through 30 April 1917, and its annual report was printed on 7 May 1917. The mail referendum showed 990 in favor of a permanent change, 611 in favor of a shorter, presumably eight-month period, 17 opposed to any change, and 33 defective ballots. The report itself appeared as a supplement in the 19 March *Current Affairs*; its context indicates it was completed in late January.

22. "Aim for Extra Hour a Day," *New York Times*, 15 November 1916, 1.

23. H.R. 19431 (2 January), H.R. 20354 (22 January), and H.R. 20499 (25 January); the Senate bill, S. 7828, was introduced on 12 January. The second session of the 64th Congress ended on 3 March 1917.

24. "'Daylight Saving' Dinner," *New York Times*, 16 January 1917, 7; "Daylight Saving Committee Dines," *New York Times*, 21 January 1917, 5; and "Labor for Daylight Plan," *New York Times*, 23 January 1917, 11.

25. "See Help for All in Saving Daylight," *New York Times*, 31 January 1917, 9; and "President Favors Saving of Daylight," ibid., 1 February 1917, 7. "Daylight Saving Advocates Will Convene To-Day," *New York Tribune*, 30 January 1917, 6; "Light Saving Advocates Begin Assault on Time," ibid., 31 January 1917, 9; and "Wilson Indorses Daylight Saving," 1 February 1917, 7.

26. Chamber of Commerce of the United States, Committee on Daylight Saving, *Report*. Also "Unanimous Consent Is Needed," *New York Times*, 2 February 1917, 7.

27. H.R. 2609, 65th Congr., 1st sess.; and S. 1854, 65th Congr., 1st sess.

28. Senate Committee on Interstate Commerce, *Daylight Saving Time and Standard Time*. Senator Joe Robinson of Arkansas, the chairman of the subcommittee, voiced his displeasure at the hundreds of letters and telegrams he had received, most of them "prompted or suggested" by the advocates. Calling them unnecessary, he refused to make them a part of the record. Chairman Robinson did include in the hearings record the 22 February 1917 report by the U.K. Home Office's Summer Time Committee on the effects of the kingdom's 1916 switch to Summer Time. Many of the British findings were, however, completely ignored by the American advocates.

29. Among them were Marcus Marks, representing the National Daylight Saving Association; longtime advocate Harold Jacoby, Rutherford Professor of Astronomy, Columbia University; Robert Garland, Chairman of the Daylight Saving Committee, United States Chamber of Commerce; A. Lincoln Filene, who cited the just-published views of the Boston Chamber of Commerce; and George Renaud, chairman, Detroit More Daylight Club.

30. Senate Committee on Interstate Commerce, *Daylight Saving Time and Standard Time*, 39, 41. The American Railway Association's resolution is dated February 1917.

31. Ibid., 5.

32. The earliest date I have found for linking increased gardening and daylight saving time directly is 30 April 1917, Boston Chamber of Commerce, "Chamber in Favor of Daylight Saving."

33. Pack, *War Gardens Victorious*, chapter 12; and Eyle, *Charles Lathrop Pack*. At the time of the commission's founding, Pack was president of the American Forestry Association.

34. O'Malley, *Keeping Watch*, 292, terms the National War Garden Commission "a lobbying organization for the makers of garden products" and accuses it of "masking itself as a government agency," of using "daylight saving to raise sales," and of invoking patriotism "to head off the competition." His views stem directly from allegations made by Representative Edward J. King on the floor of the House during the June 1919 debate to repeal the daylight-saving-time provision of the act; *Congr. Rec.*, 66th Congr., 1st sess., 18 June 1919, 58:1332–33. Both O'Malley's characterizations and King's remarks must be viewed with great skepticism, for the members of the War Garden Commission included, among others, Luther Burbank; the retired president of Harvard University, Charles W. Eliot; Congressman Myron Herrick of Ohio; and John G. Hibben, president of Princeton University; a former secretary of agriculture; and the assistant secretary of agriculture in 1917. Further, the commission was funded completely by contributions, including one of over $375,000 from Pack, its president. See Eyle, *Charles Lathrop Pack*, 142, chapter 14.

35. O'Malley, *Keeping Watch*, 273, shows postcards having the same theme sent to members of Congress, with one dated 17 May 1917. Also Prerau, *Seize the Daylight*, 86, and n39 below.

36. Senate Committee on Interstate Commerce, *Daylight Saving*, S. Rept. 46.

37. *Congr. Rec.*, 65th Congr., 1st. sess., 27 June 1917, 55:4349–53, 4355.

38. *Congr. Rec.*, 65th Congr., 1st. sess., 28 June 1917, 55:4444–47 and Appendix 234.

39. The poster campaign, which included the postcards sent to members of Congress—see n35 above—was a major propaganda effort conducted, it is believed, by the National Daylight Saving Association. Fifteen different posters, covering the period from early 1917 through October 1918, have been located in various archives. Essentially all show a link to the United Cigar Stores Company, which had retail stores in numerous communities.

40. House Committee on Interstate and Foreign Commerce, *Daylight Saving*, H. Rept. 293. One amendment removed the bill's effective date of 1 January 1918.

41. As late as 28 January Marcus Marks, in a letter to a member of Congress from Massachusetts strongly in favor of the legislation, was describing a five-month daylight saving period beginning on 28 April and ending on 29 September; see *Congr. Rec.*, 65th Congr., 2d sess., 15 March 1918, 56:3573. Most likely the lobbying came from someone associated with one of the chambers of commerce.

42. During the debates to repeal national daylight saving time, the chairman of the House committee stated that adding the two months "was made on the argument of an existing war condition requiring the application of the law for

the additional month[s] to save coal," clearly a specious line of reasoning when one considers both the weather and amount of daylight associated with April and October; see *Congr. Rec.*, 66th Congr., 1st sess., 18 June 1919, 58:1307.

43. "Move Clocks Ahead at Patriotic Rally," *New York Times*, 31 March 1918, 13.

44. U.S. Interstate Commerce Commission, "No. 10122: Standard Time Zone Investigation."

45. Ibid., 282–84.

46. For a train crossing a zone boundary, the railway company was allowed to keep the time of the departing zone into the one entered as far as the closest division or junction point. This pro forma operating exception continued for many decades. However, the roads were expected to show departure and arrival times correctly on all their advertisements, time cards, and station bulletin boards.

47. O'Malley, *Keeping Watch*, 278, 368n34, mistakenly writes that the advanced-time provision of the Standard Time Act, which he terms a "bill," expired on 27 October 1918, "which of course made it necessary to pass an entirely new bill next year [1919]." In fact, given the statute's language, no new legislation was needed.

48. S. 4982, 65th Congr., 2d sess., introduced by Senator Calder of New York, was debated and passed on 11 October 1918; see *Congr. Rec.*, 65th Congr., 2d sess., 11 October 1918, 56:11154, 11168, 11187. A similar bill, H.R. 9287, had been introduced on 26 January, prior to the passage of the daylight saving time act itself.

49. The House members introducing the bills represented districts in Illinois, Nebraska, Kansas, and South Carolina. A letter and petition from South Dakotans was also placed in the record.

50. Sources featuring the attention-garnering remarks include O'Malley, *Keeping Watch*, 280–90; Downing, *Spring Forward*, 22–33; and Prerau, *Seize the Daylight*, 103–9.

51. H.R. 3854, introduced by Representative Esch of Wisconsin, an early critic of the legislation. House Committee on Interstate and Foreign Commerce, *Hearings on Repeal of Section Three*. House Committee on Interstate and Foreign Commerce, *Repeal of Daylight Saving Law*, H. Rept. No. 42.

52. Massachusetts State Grange v. [Attorney General] Benton, *United States Supreme Court Reports*, 272 U.S. 525 (1926).

53. Doane, *Time Changes*.

54. Detroit was placed in the Eastern zone in mid-1922. The ICC hearings record for the many shifts is voluminous. For summaries of the changes, from 1918 and continuing into 1970, see U.S. Department of Transportation, Office of the Secretary, *Standard Time in the United States*. For some maps, see Bartky and Harrison, "Standard and Daylight-saving Time."

55. For examples, see Prerau, *Seize the Daylight*, and Downing, *Spring Forward*.

Epilogue

1. While at NBS in 1982, I was asked what (zone) time was being used in the Falkland Islands. The query came from the British liaison to the Army Map Service. My initial reply, that the American and British listings of zone times were different, was immediately dismissed as an ambiguity already known. My next comment—that the only reliable sources for such information were those airline companies flying into the country—was also dismissed by the remark, "Only one country's planes are flying into the Falklands right now!" Later I learned that Prime Minister Margaret Thatcher was preparing to announce the United Kingdom's Exclusion Zone in the South Atlantic surrounding the Argentine-occupied islands. Her announcement was to take effect in terms of Greenwich time, and everyone knew that some journalist was bound to ask, "What time is that in the Falklands?" For some while no one could answer this question.

2. That this particular impact was apparently unintentional is suggested the quotation given by Kristof, "In Pacific Race to Usher In Millennium, a Date Line Jog," 1:8.

3. On 10 September 1945 Tonga adopted a standard time 12 hours, 20 minutes in advance of Greenwich. On 19 October 1960 the legislation was amended by shifting to a time 13 hours in advance, and continuing the Eastern reckoning of the days. Tonga's daylight saving time, observed in 1999 and 2000, was abandoned by 2002. I thank Rhys Richards for these dates.

Not to be outdone, New Zealand extolled the virtues of seeing the Dawn of the New Age in a more comfortable tourist environment than an uninhabited island; later it would issue a commemorative stamp featuring Chatham Island and the 1 January 2000 date. And the Royal Observatory at Greenwich argued that the beginning of the millennium would not actually occur until 0000 Universal Time along the prime meridian, or *noon* on 1 January along the Date Line. While all this striving to be "first" was based on hoped-for economic benefits, it is worth noting that the world's millennium competition actually came a whole year early—for the new millennium actually began on 1 January 2001.

4. This action is the equivalent of YRDST. Subsequently advancing clocks during the summer months can be termed Double Daylight Saving Time, or DDST.

5. Since even recent maps can be out of date, I thank Tom Van Baak of Seattle for pointing out this near-foolproof approach. (Of course, countries that have multiple time zones make the problem a little more complex.)

6. Southern Hemisphere observance of advanced time is of course out of phase with observance in the Northern Hemisphere. In addition, nations located in the tropics generally do not observe advanced time.

7. The six-month period was chosen because it represented the most common observance period among DST states, which were concentrated in the Northeast, the industrial Midwest, and the West Coast. For some maps illustrating the mixed observance of DST prior to passage of the Uniform Time Act, see Bartky and Harrison, "Standard and Daylight-saving Time."

8. The hearings record is voluminous. For a summary from 1918 and continuing into 1970, see the DOT, Office of the Secretary, *Standard Time in the United States*. Subsequent changes are available from the department.

9. In 2005, Indiana enacted legislation that, starting in 2006, will place the entire state on DST. The state's action will leave only the five million inhabitants of Arizona—about 2 percent of the continental U.S. population—who do not change their clocks twice a year. See also n16 below.

10. P.L. 99-359, Fire Prevention and Control Appropriations Act of 1986.

11. Over the years many of those (counties) on the eastern side of the other three time zones have switched permanently to the time of the zone to their east.

12. Bartky and others, *Review and Technical Evaluation of the DOT Daylight Saving Time Study*; hereafter NBS Report, 1976. See also Bartky and Harrison, "Standard and Daylight-saving Time."

13. This reality is the likely reason why direct attempts, from 1976 on, to increase the DST period failed, causing its proponents to opt instead for placing riders on must-pass bills.

14. Quoted from NBS Report, 1976—see n12 above.

15. From DOT, Linda L. Lawson, "Prepared Statement," 19.

16. As an example, in Starke County, Indiana, on 11 March 2006, the sun rose at 7:06 A.M. EST. When the extended period of DST takes effect in 2007, sunrise will be at 8:06 A.M. EDT. However, if the county switches to the Central time zone, sunrise would come at 7:06 A.M. CDT. (For 4 November, the set of sunrises are 7:22 A.M. EST, 78:22 A.M. EDT, and 7:22 A.M. CDT.) Note added: As a result of the DOT's January 2006 decision, Starke County and seven other counties in Indiana are now in the Central Standard Time Zone, so on 11 March 2007 sunrise will occur at 7:06 A.M. CDT in the county.

Bibliography

This bibliography is presented in three sections, reflecting the work's major subdivisions.

I. CREATING A DATE LINE (1522–1921)

Acosta, José de. *The Natural and Moral History of the Indies.* 1590. Translated from the Spanish by Edward Grimston, 1604. Edited, with notes and an introduction, by Clements R. Markham, 1880. Reprinted as *The Hakluyt Society First Series,* vols. 60–61, New York: Burt Franklin, 1970.

Albo, Francisco. "Log-Book of Francisco Alvo or Alvaro." In *The First Voyage Round the World by Magellan,* with notes and introduction by Lord Stanley of Alderley. London: Hakluyt Society, 1874.

Baker, Emily Reynolds. *Caleb Reynolds American Seafarer.* Kingston, Ontario: Limestone Press, 2000.

[Barrow, John]. "Porter's Cruize in the Pacific Ocean." *Quarterly Review* 13 (July 1815): 374–83.

———. *The Mutiny of the* "Bounty." 1831. Oxford: Oxford University Press paperback, 1989, chapter 2.

Bartky, Ian R. "Finding and Charting the World's Time." *The Portolan* 58 (Winter 2003–4): 32–38.

Belcher, Sir Edward. *Narrative of a Voyage around the World Performed in Her Majesty's Ship Sulphur during the Years 1836–1842. . . .* 1843. Vol. 1. Reprint, Folkestone, England: Dawson's, 1970.

Bennett, Frederick Debell. *Narrative of a Whaling Voyage round the Globe, from the Year 1833 to 1836.* 2 vols. London: Richard Bentley, 1840.

Bly, Nellie. *Nelly Bly's Book: Around the World in 72 Days.* 1890. Reprint, edited by Ira Peck, Brookfield, Conn.: Twenty-First Century Books, 1998.

Boit, John. *Log of the Second Voyage of the Columbia.* 1790–93. In *Voyages of the* "Columbia" *to the Northwest coast. . . .* Edited by Frederic W. Howay. Boston: Massachusetts Historical Society, 1941. Reprint, Portland: Oregon Historical Society, [ca. 1990].

———. *Log of the Union: John Boit's Remarkable Voyage to the Northwest Coast and around the World.* 1794–96. Edited by Edmund Hayes. Portland: Oregon Historical Society, 1981.

Bougainville, Louis-Antoine de, Comte. *A Voyage round the World*. Translated from the French by John Reinhold Forster, 1772. Reprint, New York: Da Capo Press, 1967.

Bowditch, Nathaniel. In *Early American-Philippine Trade: The Journal of Nathaniel Bowditch in Manila, 1796*, edited by Thomas R. McHale and Mary C. McHale. New Haven: Yale University, South East Asia Studies, 1962.

HMS *Briton*. Captain's log, ADM 51/2184. 17 September 1814 entries. Public Record Office, Kew.

Campbell, Andrew, ed. *Impressions of Tongareva (Penrhyn Island), 1816–1901*. Suva, Fiji: Institute of Pacific Studies, University of the South Pacific, 1984.

Carletti, Francesco. *My Voyage around the World*. 1701. Translated by Herbert Weinstock. New York: Pantheon Books, 1964.

Collins, Henry. *The International Date Line*. Syracuse, N.Y.: C. W. Bardeen, 1891.

Cook, James. *The Journals of Captain Cook on His Voyages of Discovery*. Edited by J. C. Beaglehole. Hakluyt Society: Cambridge University Press, 1961.

Daily Alta California, 26 October–29 November 1867.

Dampier, William. *A New Voyage round the World*. 1697. Introduction by Sir Albert Gray. London: Argonaut Press, 1927. Reprint, New York: Da Capo Press, 1970.

Darwin, Charles. *Narrative of the Surveying Voyages of His Majesty's Ships* Adventure *and* Beagle, *between 1826 and 1836 . . . and the* Beagle's *Circumnavigation of the Globe*. Vol. 3. London: H. Colburn, 1839.

Davidson, George. "International Date Line: Clippings, Notes, Memoranda. . . ." George Davidson Papers. Bancroft Library, University of California, Berkeley.

———. Correspondence with Chief Hydrographer, March–April 1899. In NARA, Records of the U.S. Hydrographic Office. RG37, entry 32, Letters sent and received, Feb. 1885–Dec. 1901, box no. 118, folder nos. 1272 and 1472.

———. Letter to American Geographical Society regarding the Date Line, 6 August 1899, as printed in Helen M. Strong, "Universal World Time," *Geographical Review* 25 (1935): 483.

———. "There Is No International Date Line." *San Francisco Examiner*, 8 June 1901, 2.

———. *The Alaska Boundary*. San Francisco: Alaska Packers Association, 1903.

Defoe, Daniel. *Robinson Crusoe*. 1719. New York: Cosmopolitan Book Corp., 1920.

Delano, Amasa. *A Narrative of Voyages and Travels, in the Northern and Southern Hemispheres. . . .* 1817. Reprint, with an introduction by C. Hartley Grattan, New York: Praeger, [1970].

de Mofras, Eugene Duflot. "Carte détaillée des etablissements russes dans le haute Californie. . . ." [1841]. In Duflot de Mofras Collection, Bancroft Library, University of California, Berkeley.

Dick, Steven J. *Sky and Ocean Joined: The U.S. Naval Observatory, 1830–2000*. Cambridge: Cambridge University Press, 2002.

Downing, A. M. W. "Where the Day Changes." *Journal of the British Astronomical Association* 10 (1900): 176–77.

Drake, Francis. *The World Encompassed and Analogous Contemporary Documents Concerning Sir Francis Drake's Circumnavigation of the World.* 1628. With an appreciation by Richard C. Temple. London: Argonaut, 1926.

Essig, E. O. "The Russian Settlement at Ross." In *Fort Ross: California Outpost of Russian Alaska, 1812–1841,* edited by Richard A. Pierce. Kingston, Ontario: Limestone Press, 1991.

[Fitzroy, Robert], ed. *Narrative of the Surveying Voyages of His Majesty's Ships Adventure and Beagle, between 1826 and 1836 . . . and the Beagle's Circumnavigation of the Globe.* Vol. 2. London: H. Colburn, 1839.

Folger, Mayhew. "Log of the *Topaz,* 1807–8." Original held by Nantucket Historical Society, Nantucket, Mass.

Ford, Herbert. *Pitcairn: Port of Call.* Angwin, Calif.: Hawser Titles, 1996.

Gemelli Careri, John Francis. *A Voyage around the World.* Translated from Italian into English. [London: N.p., 1704].

Gunson, Neil. *Messengers of Grace: Evangelical Missionaries in the South Seas, 1797–1860.* Melbourne: Oxford University Press, 1978.

Hayden, Edward Everett. "The Present Status of the Use of Standard Time." *Publications of the U.S. Naval Observatory* 4 (1906): Appendix IV.

Hayes, Walter. *The Captain from Nantucket and the Mutiny on the Bounty.* Ann Arbor: William L. Clements Library, University of Michigan, 1996.

Heynen, William J., comp. *United States Hydrographic Office Manuscript Charts in the National Archives, 1838–1908.* NARA Special List 43. Washington, 1978.

Howse, Derek. *Greenwich Time and the Discovery of the Longitude.* Oxford: Oxford University Press, 1980.

Mykhailo Hutsuliak. *When Russia Was in America: The Alaska Boundary Treaty Negotiations, 1824–1825. . . .* Vancouver: Mitchell Press, 1971.

"The International Date Line." *School Bulletin* 4 (February 1876): 85.

Jones, G. Seneca. Letter to Representative S. G. Hilborn, 15 May 1898. In NARA, Records of the U.S. Hydrographic Office. RG37, Entry 32, Letters Sent and Received, Feb. 1885–Dec. 1901, Box no. 109, Folder no. 2324.

Kelsey, Harry. "The Gregorian Calendar in New Spain: A Problem in Sixteenth-Century Chronology." *New Mexico Historical Review* 58 (1983): 239–51.

———. *Sir Francis Drake: The Queen's Pirate.* New Haven: Yale University Press, 1998.

Kloosterman, Alfons M. J. *Discoverers of the Cook Islands and the Names They Gave.* 2d ed. [Rarotonga]: Cook Islands Library and Museum, 1976.

Kotzebue, Otto von. *Voyage of Discovery in the South Sea, and to Behring's Straits, in Search of a North-East Passage. . . .* 1815–18. Translated from the Russian. Vol. 2. London: R. Phillips, 1821, 2:220.

———. *A New Voyage Round the World in the Years 1823–1826.* Two vols. London: H. Colburn and R. Bentley, 1830.

Kristof, Nicholas D. "In Pacific Race to Usher in Millennium, a Date Line Jog." *New York Times,* 23 March 1997, 1:8.

Lamont, Roscoe. "The International Date Line." *Popular Astronomy* 29 (1921): 340–48.

Le Maire, Jacob. [Incorrectly ascribed to Willem Schouten]. *A Wonderfull Voiage Round About the World.* London, 1619. Reprint, New York: Da Capo Press, 1968.

———. *Spieghel der avstralische navigatie.* . . . (Mirror of the Australian navigation). Amsterdam: Michiel Colijn, 1622.

———. French translation. In Herrera y Tordesillas, Antonio de, *Description des Indes Occidentales.* Amsterdam: M. Colin, 1622.

Leigh-Browne, F. S. "The International Date Line." *Geographical Magazine* 14 (1942): 302–6.

Lewis, Charles Lee. "Our Navy in the Pacific and the Far East Long Ago. *United States Naval Institute Proceedings* 69 (June 1943): 857–64.

Lovett, Richard. *The History of the London Missionary Society, 1795–1895.* London: Henry Frowde, 1899.

Lutz, Cora E. "A Fourteenth-Century Argument for an International Date Line." In *Essays on Manuscripts and Rare Books.* Hamden, Conn.: Archon Books, 1975.

Malloy, Mary. *"Boston Men" on the Northwest Coast: The American Maritime Fur Trade 1788–1844.* Kingston, Ontario: Limestone Press, 1998.

Martyr, Peter [Pietro Martire d'Anghiera]. *De Nouo Orbe, or, The Historie of the West Indies . . . comprised in eight decades / written by Peter Martyr . . . whereunto the other fiue are newly added by the industrie and painefull trauaile of M. Lok Gent.* Fifth decade. London: Printed for Thomas Adams, 1612.

Miller, David Hunter. *The Alaska Treaty.* Kingston, Ontario: Limestone Press, 1981.

Mitchell, Mairin. *Elcano: The First Circumnavigator.* London: Herder, 1958.

Moerenhout, Jacques A. *Voyages aux îles du Grand océan.* 1837. Vol. 1. Reprint, Paris: Librairie d'Amérique et d'Orient, [1942]. Translated by Arthur R. Borden, Jr., as *Travels to the Islands of the Pacific Ocean,* Lanham, Md.: University Press of America, 1993.

Morrison, James. *The Journal of James Morrison, Boatswain's Mate of the Bounty . . . with an Account of the Island of Tahiti.* [London]: Golden Cockerel, 1935.

Murrell, William M. *Cruise of the Frigate Columbia Around the World.* . . . Boston: Benjamin M. Mussey, 1840.

Nature, 1869–79.

———. "The Greenwich Date." *Nature* 7 (1872): 68, 105.

New York Tribune. "Where the Day Begins. A Curiously Crooked International Date Line." *New York Tribune Illustrated Supplement,* 7 May 1899, 7.

———. "Where the Day Changes. A Curious Chronological Boundary." *New York Tribune Illustrated Supplement,* 27 August 1899, 15.

Newbury, Colin. *Tahiti Nui: Change and Survival in French Polynesia 1767–1945.* Honolulu: University Press of Hawai'i, 1980.

Ohio Education, 1878–81.

———. "Where Does Sunday Begin?" *Ohio Education* 28 (May 1879): 137.

Paullin, Charles O. *American Voyages to the Orient 1690–1865*. Annapolis, Md.: U.S. Naval Institute, [1971].

Pigafetta, Antonio. "Pigafetta's Account of Magellan's Voyage." In *The First Voyage Round the World by Magellan*, with notes and introduction by Lord Stanley of Alderley. London: Hakluyt Society, 1874.

———. In *Magellan's Voyage around the World by Antonio Pigafetta*, with English translation . . . by James Alexander Robertson. Vol. 2. Cleveland: Arthur A. Clark, 1906.

Pipon, Capt. Philip. "Narrative Relating to the Pitcairn Islanders." September 1814. In "Papers Concerning the Discovery of Pitcairn Island and the Mutineers of HMS *Bounty*, 1808–1809, 1813–1815, [1845]," Sir Joseph Banks Electronic Archives, Series 71: 05–06, State Library of New South Wales, Australia.

Potomac, U.S. Frigate. Deck log. In NARA, Records of the Department of the Navy. RG24, Logbooks of U.S. Navy Ships. 24 October 1832 entries.

[Pretty, Francis]. *The Voyages of Capt. Thomas Cavendish, the Second English Navigator Who Sailed Round the World*, [1586–88]. In *An Historical Account of All the Voyages Round the World Performed by English Navigators*. London: Printed for F. Newbery, 1774, 1:159–278.

Richards, E. G. *Mapping Time: The Calendar and Its History*. Oxford: Oxford University Press paperback, 1999.

Rodgerson, J., and G. Stallworthy, 25 August 1835, to Directors, London. In Council for World Mission Archives, London Missionary Society, South Seas (1796–1840), Box 10, Folder 3.

Rogers, Woodes. *A Cruising Voyage Round the World*. 1712. With an introduction and notes by G. E. Manwaring. New York: Longmans, Green, 1929.

Ross, James Clark. *A Voyage of Discovery and Research in the South and Antarctic Regions, During the Years 1839–43*. Vol. 2. London: John Murray, 1847.

[Rudmose-Brown, R. N.]. "Wrangel [*sic*] Island." *Encyclopædia Britannica*. Chicago: Encyclopædia Britannica, 1943.

Schurz, William L. *The Manila Galleon*. New York: E. P. Dutton, 1939, Dutton paperback, 1959.

Selga, Miguel. "The Philippines in the West and in the East." *Publications of the Manila Observatory* 2 (1931): 4–15.

Shillibeer, John. *A Narrative of the Briton's Voyage to Pitcairn's Island*. London: Law and Whittaker, 1817 and 1818.

Smith, Benjamin E. "Where a Day Is Lost or Gained." *Century Magazine* 58 (September 1899): 742–45.

Steele, Richard. *The Englishman: A Political Journal by Richard Steele*, no. 26 (3 December 1713). Edited by Rae Blanchard. Oxford: Clarendon Press, 1955, 106–9.

Stevenson, Margaret. *Letters from Samoa, 1891–1895*. New York: Charles Scribner's Sons, 1906.

Stewart, Charles S. *A Visit to the South Seas in the U.S. Ship Vincennes, During the Years 1829 and 1830*. . . . Vol. 2. New York: J. P. Haven, 1831, 2:23.

Stieler, Adolf. *Stieler's Handatlas über alle Theile der Erde und über das Welt-gebäude.* Gotha: J. Perthes, [1892].

Stommel, Henry. *Lost Islands.* Vancouver: University of British Columbia Press, 1984.

Strong, Helen M. "Universal World Time." *Geographical Review* 25 (1935): 479–84.

HMS *Tagus.* Captain's log, ADM 51/2873; and Master's log, ADM 52/4629. 17 September 1814 entries. Public Record Office, Kew.

Thomas, Pascoe. *A True and Impartial Journal of a Voyage to the South-Seas, and Round the Globe, In His Majesty's Ship the Centurion Under the Command of Commodore George Anson.* London: S. Birt, 1745.

Topliff, Samuel. "Pitcairn's Island." *New England Galaxy* 4 (12 January 1821): 53–54.

United Kingdom. Admiralty. Hydrographic Department. "Notes on the History of the Date or Calendar Line," November 1921. *New Zealand Journal of Science and Technology* 11 (April 1930): 385–88.

United Kingdom. United Kingdom Hydrographic Office. Archives. "History of the Date Line." File H6948/1921.

United States. President. Message. *Russian America,* 40th Congr., 2d sess., 1868. H. Ex. Doc. 177.

Vancouver, George. *Voyage of Discovery to the North Pacific Ocean and Round the World.* London, 1798. Vol. 3. Reprint, New York: Da Capo Press, 1968.

Veniaminov, Ioann. *Journals of the Priest Ioann Veniaminov in Alaska, 1823 to 1836.* Translation of the Russian original by Jerome Kisslinger. Fairbanks: University of Alaska Press, 1993.

Verne, Jules. *Around the World in 80 Days.* 1872. Mahwah, N.J.: Watermill Press, 1981.

Vincennes, U.S. Ship. Deck log. In NARA, Records of the Department of the Navy. RG24, Logbooks of U.S. Navy Ships. 30 December 1829, 1 January 1830 entries.

Warriner, Francis. *Cruise of the United States Frigate Potomac Round the World During the Years 1831–1834.* Boston: Crocker & Brewster, 1835, 219–20.

Wilkes, Charles. *Narrative of the United States Exploring Expedition . . . 1838–1842.* Vols. 2, 3, 5. Philadelphia: Lea and Blanchard, 1845, 2:159, 3:367, 5:267.

II. CAMPAIGNING FOR UNIFORM TIME (1870–1925)

Abbe, Cleveland. Letter to Sandford Fleming, 24 May 1895. In Fleming Papers, Library and Archives Canada, vol. 1.

———. "Report of Committee on Standard Time." *Proceedings of the American Metrological Society* 2 (1880): 17–45.

Abbe, Truman. *Professor Abbe . . . and the Isobars; the story of Cleveland Abbe, America's first weatherman.* New York: Vantage Press, 1955.

Airy, George B., to The Secretary of State for the Colonies, 18 June 1879. In Fleming, *Universal or Cosmic Time*, suppl. papers, 32–34.

Aitken, Hugh G. J. *Syntony and Spark: The Origins of Radio.* Princeton: Princeton University Press, Princeton paperback, 1985.

Albrecht, Th. "Über die Verwendbarkeit der drahtlosen Telegraphie bei Lägenbestimmungen." *Astronomische Nachrichten* 166 (5 November 1904): cols. 337–44.

American Association for the Advancement of Science. *Proceedings of . . . the Thirty-second Meeting, August 1883.* Salem, Mass.: The Association, 1884.

American Metrological Society, Proceedings of, 1873–88.

American Society of Civil Engineers, Proceedings of, 1881–85.

Astronomical and Physical Society of Toronto. Joint Committee. Circular-Letter, "Proposed Change in Reckoning the Astronomical Day," 21 April 1893; and First Report, 20 April 1893. *Transactions of the Astronomical and Physical Society of Toronto, for the year 1892* (1893): Appendix iv.

―――. Second Report, "Unification of the Astronomical, Civil, and Nautical Days," 10 May 1894. *Transactions of the Astronomical and Physical Society of Toronto, for the Year 1894* (1895): Appendix. Reprinted in *Proceedings and Transactions of the Royal Society of Canada,* 2d ser., 2 (1896), Appendix, A13–21.

―――. Third Report, "Unification of Astronomical, Civil and Nautical Time," 26 September 1895. *Transactions of the Astronomical and Physical Society of Toronto, for the year 1895* (1896): 95–100. Reprinted in *Proceedings and Transactions of the Royal Society of Canada,* 2d ser., 2 (1896), Appendix, A23–28.

―――. *Transactions of the Astronomical and Physical Society of Toronto, for the Year 1897* (1898): 51–55, 61–70.

Astronomische Gesellschaft. *Vierteljahrsschrift der Astronomischen Gesellschaft* 20 (1885): 216, 227–31.

Audoin, Claude, and Bernard Guinot. *The Measurement of Time: Time, Frequency and the Atomic Clock.* Translated by Stephen Lyle. Cambridge: Cambridge University Press, 2001.

Baillaud, Benjamin, and [Edmée] Chandon. "Rapport relatif aux signals horaires émis de l'Observatoire de Paris par le post radiotélégraphique de la tour Eiffel. . . ." Chapter 1, History. *Bulletin Astronomique,* 2d ser., 2 (1922): 129–37.

Barnard, F. A. P. "Adoption d'un maître méridien." In International Geographical Congress, IGC 3, 2:7–9.

―――. Letter to Secretary of State, 22 September 1884. In NARA, Records of the Department of State. RG59, Miscellaneous Letters of the Department of State, M179, roll 660, counter index 0536–38.

Bartky, Ian R. "The Invention of Railroad Time." *Railroad History* 148 (Spring 1983): 13–22.

―――. "The Adoption of Standard Time." *Technology and Culture* 30 (1989): 25–56.

———. *Selling the True Time: Nineteenth-Century Timekeeping in America.* Stanford: Stanford University Press, 2000.

———. "Sandford Fleming's First Essays on Time." (submitted for publication).

Belknap, George E., and Simon Newcomb, to Chief of Bureau of Navigation, 31 March 1886. Quoted in United Kingdom, Science and Art Department, *Thirty-fourth Report* [for 1886], Appendix A, 12–15.

Berton, Pierre. *The Impossible Railway: The Building of the Canadian Pacific.* New York: Alfred A. Knopf, 1972.

Bigourdan, Guillaume. "Sur la distribution de l'heure à distance, au moyen de la télégraphie électrique sans fil." *Comptes rendus* 138 (1904): 1657–59.

———. "Sur l'application de la télégraphie sans fil à l'amélioration des avertissements météorologiques." *Comptes rendus* 146 (27 April 1908): 885–87.

———. "Le jour et ses divisions (B.1–50), Les fuseaux horaires (B.51–68), Transmission de l'heure (B.69–82), [et] La Conférence internationale de l'heure" (B.83–104). In Bureau des Longitudes, *Annuaire pour l'an 1914.*

Blaise, Clark. *Time Lord: Sir Sandford Fleming and the Creation of Standard Time.* London: Weidenfeld & Nicolson, 2000.

Bouquet de la Grye, Anatole. "Note sur l'adoption d'une heure légale en France." *Comptes rendus* 107 (1888): 56.

———. "L'heure nationale." *Revue Scientifique*, 4th ser., 9 (7 May 1898): 579–81.

———. "Détermination de l'heure, sur terre et sur mer, à l'aide de la télégraphie sans fil." *Comptes rendus* 146 (30 March 1908): 671–73.

Bourinot, John George. "The Unification of Time at Sea." *Proceedings and Transactions of the Royal Society of Canada*, 2d ser., 2 (1896): A3–6.

British Association for the Advancement of Science. *Journal of Sectional Proceedings, no. 6, issued Tuesday Morning, September 12, 1876.* In BAAS Archives, Bodleian Library, Oxford University.

———. "Notices and Abstracts of Miscellaneous Communications to the Sections." In *Report of the Forty-sixth Meeting . . .* [for 1876], xiv, 182.

———. *Journal of Sectional Proceedings, no. 6, issued Tuesday Morning, August 20, 1878.* In BAAS Archives, Bodleian Library, Oxford University.

———. *Report of the Forty-eighth Meeting . . .* [for 1878].

———. *Report of the Fifty-ninth Meeting . . .* [for 1889], 49.

———. *Report of the Sixty-seventh Meeting . . .* [for 1897], 550–51, 720.

———. Report of the Council, 7 September 1898. In *Report of the Sixty-eighth Meeting . . .* [for 1898], lxxxiv.

British Astronomical Association, *Journal of the,* 1890–1908.

———. "Unification of Time at the Cape." *Journal of the British Astronomical Association* 2 (May 1892): 420.

———. "Universal Time." *Journal of the British Astronomical Association* 3 (March 1893): 240.

———. "Change of Time in Australia." *Journal of the British Astronomical Association* 5 (1895): 310.

———. "Uniform Time for South Africa." *Journal of the British Astronomical Association* 5 (1895): 311.

————. "A Standard Time for India." *Journal of the British Astronomical Association* 15 (1905): 389.

Brown, Lloyd A. *The Story of Maps.* 1949. Reprint, New York: Dover, 1979.

Bucholz, Arden. *Moltke, Schlieffen, and Prussian War Planning.* 1991. Providence, R.I.: Berg, paperback edition, 1993.

Bureau des Longitudes. "Unification de l'heure en France, en Algérie et en Tunisia." *Annuaire pour l'an 1889,* 779–80.

————. "Congrès international des éphémérides astronomiques." 1911. *Annales du Bureau des Longitudes* 9 (1913): A.1–51.

————. "Conférence internationale de l'Heure." 1912. *Annales du Bureau des Longitudes* 9 (1913): D.1–286.

————. "Longitude." *Annuaire pour l'an 1916,* 156–57.

Burpee, Lawrence J. *Sandford Fleming, Empire Builder.* Oxford: Oxford University Press, 1915.

Canada. House of Commons. "Return to an Order of the House of Commons, dated the 15th May, 1891: For Copies of All Letters, Communications and Reports in the Possession of the Government, Relating to the Fixing of a Standard of Time and the Legalization Thereof." *Sessional Papers,* 1891, no. 44.

Canadian Institute, (Royal). Council. "Memorial to Marquis of Lorne, Governor-General of Canada." [1879]. In Fleming, *Papers on Time-Reckoning and the Selection of a Prime Meridian to be Common to all Nations,* 5–7; and Fleming, *Universal or Cosmic Time,* suppl. papers, 27–28.

————. Joint Committee. Circular-Letter, "Proposed Change in Reckoning the Astronomical Day," 21 April 1893; and First Report, 20 April 1893. *Transactions of the Canadian Institute* 3 (1891–92): 307–16.

Caspari, Édouard. *Rapport fait au nom de la Commission de l'unification des longitudes et des heures.* [Paris]: Ministère de l'Instruction Publique et des Beaux-Arts, Imprimerie nationale, August 1884.

Cathenod, H. "L'heure fuselaire." *La Nature* 39 (25 February 1911): 211–16.

Chasseaud, Peter. *Artillery's Astrologers: A History of British Survey and Mapping on the Western Front 1914–1918.* Lewes, East Sussex: Mapbooks, 1999.

Christie, William H. M., to Secretary, Science and Art Department, 20 April 1885. In United Kingdom, Science and Art Department, *Thirty-third Report* [for 1885], Appendix A, 27–28.

————. "Universal Time." *Royal Institution of Great Britain Proceedings* 11 (1884–86): 387–94. Also published as "Universal or World Time," *Nature* 33 (1 April 1886): 521–23, with last three paragraphs omitted.

Ciel and Terre, 1881–92.

————. "L'heure de Greenwich." *Ciel et Terre* 13 (April 1892): 93–94.

Comité international des Poids et Mesures. *Procès-verbaux des séances de 1884.*

Cook, James. *Log-Book of the "Resolution."* In J. C. Beaglehole (ed.), *The Journals of Captain Cook on His Voyages of Discovery.* Hakluyt Society: Cambridge University Press, 1961, 2:cxli, n1.

Creet, Mario. "Sandford Fleming and Universal Time." *Scientia Canadensis* 14 (1990): 66–89.

Crommelin, A. C. D. "Time at Sea and the Astronomical Day." *Nature* 101 (25 April 1918): 146–47.

———. "The New System of Time-Keeping at Sea." *Nature* 101 (20 June 1918): 307–8.

Cutts, R. D. "Communication on the Action of the International Geodetic Association." *Bulletin of the Philosophical Society of Washington* 6 (1884): 106–10.

d'Abbadie, Antoine. "Choix d'un premier méridien." *Revue d'Astronomie Populaire* 3 (1884): 408–9.

de Beaumont, Henry Bouthillier. "Le méridien unique." *L'Exploration* 1 (1877): 131–32.

———. "Choix d'un méridien initial." *L'Exploration* 7 (12 January 1879): 132–36.

———. "Note [d'un méridien initial]." *Le Globe* 18 (1879): 202–8, with map.

———. *Choix d'un méridien initial unique.* Geneva: Libraire Desrogis, [June] 1880.

———. "De la projection en cartographie. . . ." *Le Globe* 27 (1888): 1–26.

———. "Utilité de l'adoption d'un méridien initial et d'une heure universelle." In International Geographical Congress, IGC 4, 1:205–9.

De Busschere, Louis. "L'unification des heures et son application en Belgique." *Bulletin de la Société royale belge de géographie* 14 (1890): 253–300, with two maps.

———. "L'unification des heures." *Ciel et Terre* 11 (1 September 1890): 301–12. Extracts from and additions to "L'unification des heures et son application en Belgique."

———. "L'unification des heures, situation actuelle." *Bulletin de la Société royale belge de géographie* 15 (1891): 297–349.

———. "L'unification des heures en Europe." *Bulletin de la Société royale belge de géographie* 16 (April 1892): 196–212.

de Claparède, A. "Henry Bouthillier de Beaumont." *Le Globe* 37 (1898): 1–14.

de Nordling, Wilhelm. "L'unification des heures." *Bulletin de la Société de Géographie*, 7th ser., 11 (1890): 111–37.

———. "Le projet de loi sur l'unification horaire." *Le Génie Civil* 17 (21 June 1890): 115–17.

———. "Les derniers progrès de l'unification de l'heure." *Compte rendu des séances de la Société de géographie et de la commission centrale* (1893): 204–5; English-language summary, *Nature* 48 (3 August 1893): 330–31.

de Rochelle, Roux. "Mémoire sur la fixation d'un premier méridien." *Bulletin de la Société de Géographie*, 4th ser., 3 (March 1845): 145–53.

Dick, Steven J. *Sky and Ocean Joined: The U.S. Naval Observatory, 1830–2000.* Cambridge: Cambridge University Press, 2002.

Dowd, Charles F. *System of National Time and Its Application . . . to the National Railway Time-Table. . . .* Albany, N.Y.: Weed, Parsons, 1870.

———. "Origin and Early History of the New System of National Time." *Proceedings of the American Metrological Society* 4 (1883): 90–101.

———. Railroad section map. Reprinted in *Railroad History* 148 (Spring 1983): front cover.

Dreyer, J. L. E., and H. H. Turner, eds. *History of the Royal Astronomical Society 1820–1920*. London: Royal Astronomical Society, 1923.

Dyson, F. W., and H. H. Turner. "The Commencement of the Astronomical Day," 19 July 1917. *Observatory* 40 (August 1917): 301–2, (September 1917): 324–27, (October 1917): 377.

Ehrenberg, Ralph E. "'Up in the Air in More Ways than One': The Emergence of Aviation Cartography in the United States." In *Cartographies of Travel and Navigation*, edited by James R. Ackerman. Chicago: University of Chicago Press, 2006.

Electrician. "Wireless Telegraphy." *Electrician (London)* 41 (14 October 1898): 815.

Ellis, William. "The Prime Meridian Conference." *Nature* 31 (6 November 1884): 7–10.

Faye, Hervé. "Sur l'heure universelle proposée par la Conférence de Rome." *Comptes rendus* 97 (1883): 1234–39.

Ferrié, Gustave. [Report on the Transmission of Time Signals]. 24 February 1909. Summarized in Baillaud and Chandon, "Rapport relatif aux signals horaires," 130.

———. "Sur quelques nouvelles applications de la télégraphie sans fil." *Journal de Physique*, 5th ser., 1 (1911): 181–88.

———. [Report on the Time Differences between Two Wireless Time Signals]. 1911 or 1912. Noted in Baillaud and Chandon, "Rapport relatif aux signals horaires," 135.

Fleming, Sandford. Sandford Fleming Papers. Library and Archives Canada, Ottawa.

———. *Terrestrial Time*. London: E. S. Boot, [ca. March 1878].

———. *Temps terrestre: Mémoire*. Paris: E. Lacroix, [August] 1878.

———. *Uniform Non-local Time (Terrestrial Time)*. [London and North America]: N.p., [ca. November 1878].

———. "Time-Reckoning." *Proceedings of the Canadian Institute*, n.s., 1 (1879): 97–137.

———. "Longitude and Time-Reckoning." *Proceedings of the Canadian Institute*, n.s., 1 (1879): 138–49.

———. *Papers on Time-Reckoning and the Selection of a Prime Meridian to be Common to all Nations*. Toronto: Copp, Clark, 1879.

———. "Adoption d'un maître méridien." In International Geographical Congress, IGC 3, 2:10–17. Reprinted in English as "On the Regulation of Time and the Adoption of a Prime Meridian," in Fleming, *Universal or Cosmic Time*, 56–62.

———. "On Uniform Standard Time, For Railways, Telegraphs and Civil Purposes Generally." *Transactions of the American Society of Civil Engineers* 10 (December 1881): 387–92.

———. Chairman's Reports. *Proceedings of the American Society of Civil Engineers* 8 (1882): 4–6, 73–75; 9 (1883): 2–4, 108–10; 10 (1884): 37–39, 76–81, with discussions following. Additional annual reports, with the final one presented on 13 January 1900.

———. *Letter to the President of the American Society [Association] for the Advancement of Science on the Subject of Standard Time for the United States of America, Canada and Mexico.* N.p., [August 1882].

———. "Cosmopolitan Scheme for Regulating the Time." *Proceedings of the American Metrological Society* 2 (1882): 294–301, with four figures facing 332.

———. *Universal or Cosmic Time.* Toronto: Coop, Clark, 1885.

———. "Report on the Washington International [Meridian] Conference." In *Universal or Cosmic Time,* suppl. papers, 67–82.

———. "Memorandum on the Movement for Reckoning Time on a Scientific Basis." 20 November 1889. *Transactions of the Canadian Institute* 1 (1889–90): 227–38. Also printed in United Kingdom, Science and Art Department, *Thirty-eighth Report* [for 1890], Appendix A, 16–24; and reprinted in Canada, House of Commons, *Sessional Papers,* 1891, no. 44, 3–11, with map.

———. "Bibliography." *Proceedings and Transactions of the Royal Society of Canada for the Year 1894* 12 (1895): 33–35.

———. Letter to Cleveland Abbe, 21 May 1895. In Fleming Papers, Library and Archives Canada, vol. 1.

———. "Unification of Time at Sea." *Times* (London), 9 January 1897, 12.

Foerster, Wilhelm. "Ueber die von der Conferenz zu Washington proponirte Veränderung des astronomischen Tagesanfanges." *Astronomische Nachrichten* no. 2643 (21 January 1885): cols. 33–38.

Forel, F. A. "L'unification de l'heure; l'heure nationale." *Revue d'Astronomie Populaire* 7 (September 1888): 327–32.

France. *Journal officiel de la Republique Française.* 29 and 30 March 1917.

———. Service Hydrographique de la Marine. *Guide pour la lecture des Cartes marines étrangères.* Paris: Imprimerie nationale, 1917.

———. "Limites des fuseaux horaires en mer." *Avis aux navigateurs,* nos. 548–52. 16 April 1917.

———. Carte no. 7, "Planisphère des fuseaux horaires," 1:72,000,000. June 1917.

Franklin, Samuel R. *Memories of a rear-admiral who has served for more than half a century in the Navy of the United States.* New York: Harper & Brothers, 1898.

Galison, Peter. *Einstein's Clocks, Poincaré's Maps: Empires of Time.* New York: W. W. Norton, 2003.

Gautier, Étienne. "Est-il opportun de changer le Mode de computer le Jour en Astronomie?" *Observatory* 8 (December 1885): 420–24.

Geographical Society of Hamburg. *Mittheilungen der Geographischen Gesellschaft in Hamburg* 2 (1882–83): 294–95.

Germain, Adrien. "Le premier méridien et la Connaissance des temps." *Bulletin de la Société de Géographie,* 6th ser., 9 (May 1875): 504–21.

Gill, David. Letter to Secretary of the Admiralty, 2 February 1892. In United Kingdom, Science and Art Department, *Thirty-ninth Report* [for 1891], Appendix A, 25.

Green, Lorne. *Chief Engineer: Life of a Nation-Builder, Sandford Fleming.* Toronto: Dundurn Press, 1993.

Greenwood, Nelson. "The Unification of Time." *Proceedings and Transactions of the Royal Society of Canada,* 2d ser., 2 (1896), Appendix: A11, A29–41.

Grubb, Sir Howard. "Proposal for the Utilisation . . . of Wireless Telegraphy for the Control of Public and Other Clocks." *Scientific Proceedings of the Royal Dublin Society* 9 (1899): 46–49.

Guinot, Bernard. "Le Bureau International de l'Heure de 1911 à 1964: Le temps astronomique et la naissance du temps atomique." In Paris Observatory, *Le Bureau international de l'heure: 75 ans au service de l'heure universelle,* edited by Martine M. Feissel and Suzanne Débarbat. Meudon, France: Observatoire de Paris-Bureau des Longitudes, 1992.

Guyou, Émile. "Détermination des longitudes en mer par la télégraphie sans fil." *Comptes rendus* 146 (15 April 1908): 800–801.

Haag, Heinrich. *Die Geschichte des Nullmeridians.* Leipzig: Otto Wigand, 1913, with map.

Hall, Asaph. "Reports of the National [A]cademy of [S]ciences." *Science* 7 (1886): 286.

———. "Proposed Change in the Astronomical Day." *Observatory* 9 (1886): 161.

Hall, Asaph, Jr. Manuscript Trip Report. [November 1912]. In U.S. Naval Observatory Archives, drawer BF, Asaph Hall, Jr., folder, 7–8.

Hazen, Wm. B. "The Adoption of a Prime Meridian." In International Geographical Congress, IGC 3, 2:18–19.

Herschel, J. F. W. *Outlines of Astronomy.* 1st ed. London: Longman, Brown, Green, and Longmans, 1849.

Hirsch, A., and Th. v. Oppolzer, eds. *Unification des longitudes par l'adoption d'un Méridien initial unique et introduction d'une heure universelle.* [Berlin, 1884]. Extracted from International Geodetic Association, *Comptes-rendus des séances de la Septième Conférence Générale de l'Association géodésique internationale.*

Holden, E. S. Remarks, 10 August 1885. Included in U.S. Senate, Committee on Naval Affairs, *Letter from the Secretary of the Navy Transmitting . . . Report of the National Academy of Sciences . . . ,* 28–30.

———. *Memorials of William Cranch Bond And of His Son George Phillips Bond.* San Francisco: C. A. Murdock, 1897.

Holdich, T. H. "The Prime Meridian." *Proceedings of the Royal Geographical Society* 13 (1891): 615–16.

Howse, Derek. *Greenwich Time and the Longitude.* London: Philip Wilson, 1997. (A revised edition of *Greenwich Time and the Discovery of the Longitude.* Oxford: Oxford University Press, 1980.)

Hutchinson, D. L. "Wireless Time Signals from the St. John Observatory of the Canadian Meteorological Service." *Proceedings and Transactions of the Royal Society of Canada,* 3rd ser., 2 (1908): 153–54.

International Geodetic Association. *Comptes-rendus des séances de la Septième*

Conférence Générale de l'Association géodésique internationale. Procès-verbaux. Berlin: Georg Reimer, 1884.

————. "Rapport de la Commission permanente pour l'année 1883." In *Comptes-rendus des séances,* 127–37.

————. *Extrait des Comptes-rendus des séances de la Septième Conférence Générale de l'Association géodésique internationale.* [Berlin]: Bureau central de l'Association géodésique internationale, [1884].

International Geographical Congress, 1871–1915.

————. *Compte-rendu du Congrès des Sciences Géographiques, Cosmographies et Commerciales tenu à Anvers du 14 au 22 août 1871.* 2 vols. Anvers, 1872. Cited in notes as IGC 1.

————. *Congrès International des Sciences Géographiques tenu à Paris du 1er au 11 Août 1875; Compte rendu des séances.* 2 vols. Paris: Société de Géographie, 1880. Cited in notes as IGC 2.

————. *Terzo Congresso Geografico Internazionale tenuto a Venezia dal 15 al 22 Settembre 1881.* Vol. 1. *Notizie e Rendiconti.* Rome: Società Geografica Italiana, 1882. Vol. 2. *Comunicazioni e Memorie.* Rome: Società Geografica Italiana, 1884. Kraus reprint, 1972. Cited in notes as IGC 3.

————. *IVe Congrès International des Science Géographiques tenu à Paris en 1889.* Vol. 1. *Compte rendu publié par le Secrétariat général du Congrès.* Paris: Bibliothèque des Annales Économiques Société d'Éditions Scientifiques, 1890. Cited in notes as IGC 4.

International Map Committee. *Resolutions and Proceedings of the International Committee assembled in London, November, 1909.* London: His Majesty's Stationery Office, 1910.

International Meteorological Congress, 1873–93.

————. Deuxième Congrès International de Météorologie. *Berichte des Professor Dr. C. Bruhns über de Fragen . . . 33 . . . des Programms für den Meteorologen-Congress in Rom. 1879.* Rome: Imprimerie Héritiers Botta, 1879, 17–24.

————. *Procès-verbeaux et appendices.* Rome: Imprimerie Héritiers Botta, 1879, 16–18, 153.

————. *Rapports sur les questions du programme.* Rome: Imprimerie Héritiers Botta, 1879, 195–97.

International Telegraph Conference. *Regulations,* 1875–1915.

————. *Document de la [1890] Conférence Télégraphique Internationale de Paris publiés par le Bureau International des Administrations télégraphiques.* Berne: Rieder & Simmen, 1891, 328, 635.

Jackson, J. R. "Chartography." In *Manual of Geographical Science.* Vol. 1. London: John W. Parker and Son, 1855.

James, Colonel Sir Henry. *On the Rectangular Tangential Projection of the Sphere and Spheroid . . . for a Map of the World on the Scale of Ten Miles to an Inch.* Southhampton: Ordnance Survey Office, 1868.

Janssen, Jules. "Sur le Congrès de Washington. . . ." *Comptes rendus* 100 (1885): 706–26.

————. "Notice sur le Méridien et l'Heure universels." In Bureau des Longitudes, *Annuaire pour l'an 1886*, 835–81.

Kikuchi, D. "Time Reform in Japan." *Nature* 34 (1886): 469.

La Nature, 1910–19.

————. "L'heure légale française." *La Nature* 26 (2 April 1898): 287.

Lallemand, Charles. "L'unification internationale des heures et le système des fuseaux horaires." *Revue Scientific*, 4th ser., 7 (3 April 1897): 419–25.

————. "L'heure légale en France et les fuseaux horaires." *Revue Scientific*, 4th ser., 9 (16 April 1898): 491–97.

————. "Sur un projet de Carte internationale et de Repères aéronautiques." *Comptes rendus* 152 (29 May 1911): 1439–46.

————. "Cartes et repères aéronautiques" (International Air Map and Aeronautical Marks). Presented at the Fifth International Aeronautical Conference, Turin, October 1911. In *Revue générale des sciences pures et appliquées* 22 (30 July 1911): 557–62. English translation, *Geographical Journal* 38 (1911): 469–83; reprinted in *Annual Report of the Board of Regents of the Smithsonian Institution for 1911*, 295–302.

————. "Projet d'organisation d'un service international de l'heure." In "Conférence internationale d'Heure," *Annales du Bureau des Longitudes* 9 (1913): D.261–68.

————. "L'heure à bord des navires." *Comptes rendus* 164 (2 April 1917): 544–45.

————. "À propos de l'extension, à la mer, du régime des fuseaux horaires." *Comptes rendus* 165 (23 July 1917): 131–33.

Lallemand, Charles, and Joseph Renaud. "Substitution du temps civil au temps astronomique dans les Éphémérides nautiques." *Comptes rendus* 166 (11 March 1918): 401–2, 708; and *L'Astronomie* 32 (1918): 177–79.

Lambert, Armand. "Le Bureau International de l'Heure: Son rôle, son fonctionnement," 2. Extract from Bureau des Longitudes, *Annuaire pour l'an 1940*.

Levallois, J. J. "The History of the International Association of Geodesy." *Bulletin géodésique* 54 (1980): 259–313.

————. *Mesurer la Terre: 300 ans de géodésie française*. Paris: Presses de l'École nationale des ponts et chaussées, 1988.

Malin, Stuart R. "The International Prime Meridian Conference, Washington, October 1984 [1884]." *Journal of Navigation* 38 (1985): 203–6.

Mayall, R. Newton. "The Inventor of Standard Time." *Popular Astronomy* 50 (1942): 204–9.

McCusker, John J. "How Much Is That in Real Money? A Historical Price Index for Use as a Deflator of Money Values in the Economy of the United States." *Proceedings of the American Antiquarian Society* 101 (October 1991): 297–373.

Moitoret, V. A. *The International Hydrographic Bureau: 50 Years of Progress*. Monaco: International Hydrographic Organization, 1971.

Moltke, Helmuth von. *Gesammelte Schriften und Denkwürdigkeiten des General-Feldmarschalls Grafen Helmuth von Moltke*. Berlin: Ernst Siegfried

Mittler und Sohn, 1892, 7:38–43. French translation by Louis De Busschere in "L'unification des heures, situation actuelle," 322–26. English translation, not in its entirety, by Sandford Fleming, printed in Dominion of Canada, House of Commons, *Sessional Paper no. 44* (1891), 25–27, and in *Proceedings of the American Society of Civil Engineers* 18 (1892): 58–60.

Mueller, Ivan I. "125 Years of International Cooperation in Geodesy." In *International Association of Geodesy Symposium 102*. New York: Springer-Verlag, 1990, 421–32.

National Academy of Sciences. "The Astronomical Day." In U.S. Senate, Committee on Naval Affairs, *Letter from the Secretary of the Navy Transmitting . . . Report of the National Academy of Sciences upon the Proposed New Naval Observatory*, 49th Congr., 1st sess., 10 February 1886. S. Ex. Doc. 67, 2–4. Reprinted in United Kingdom, Science and Art Department, *Thirty-fourth Report* [for 1886], Appendix A, 19–22.

National Archives and Records Administration. *List of Logbooks of U.S. Navy Ships, Stations . . . 1801–1941*. Special List 44. Washington, 1978, 1n1.

Nature, 1874–1900, 1908–18.

———. "The International Congress and Exhibition of Geography." *Nature* 12 (12 August 1875): 293–94.

———. "Many Proposals Have Been Made." *Nature* 19 (16 January 1879): 247.

———. "The International Conference." *Nature* 29 (14 February 1884): 367.

———. "Our Future Watches and Clocks." *Nature* 31 (13 November 1884): 36–37; (27 November 1884): 80; (11 December 1884): 128; (1 January 1885): 201–2; (8 January 1885): 217; (15 January 1885): 241–42; (5 February 1885): 317.

———. "The Question of Civil and Astronomical Time." *Nature* 32 (16 July 1885): 245–47.

———. "Time Standards of Europe." *Nature* 46 (23 June 1892): 175.

———. "Universal Time." *Nature* 47 (9 March 1893): 451.

———. "France and International Time." *Nature* 48 (3 August 1893): 330–31.

———. "Standard Time in Australia." *Nature* 51 (28 March 1895): 516.

———. "Physics at the British Association." [Status regarding changing the Astronomical Day]. *Nature* 56 (9 September 1897): 460.

———. "France and International Time." *Nature* 58 (19 May 1898): 60.

Newcomb, Simon. "On the Proposed Change of the Astronomical Day." *Monthly Notices of the Royal Astronomical Society* 45 (9 January 1885): 122–23.

———. *The Reminiscences of an Astronomer*. Boston and New York: Houghton, Mifflin, 1903.

Norie, J. W. *A New and Complete Epitome of Practical Navigation: Containing All Necessary Instructions for Keeping a Ship's Reckoning at Sea*. 4th ed. London: J. W. Norie, 1816.

Normand, A. "Sur le réglage des montres à la mer par la télégraphie sans fil." *Comptes rendus* 139 (1904): 118.

Observatory, 1885–86, 1908–19.

———. "The Proposed Change in the Astronomical Day." *Observatory* 8 (March 1885): 81–84.

———. [Telegram to *Titanic*]. *Observatory* 35 (June 1912): 247–48.

———. "An International Time Conference." *Observatory* 35 (November 1912): 411–12.

Oppolzer, Theodor Ritter von. "On the Proposed Change of the Astronomical Day." *Monthly Notices of the Royal Astronomical Society* 45 (13 March 1885): 295–98.

———. *Canon der Finsternisse.* 1887. Translated by Owen Gingerich, with a preface by Donald H. Menzel and Owen Gingerich. New York: Dover, 1962.

Paris Observatory. *Le Bureau international de l'heure: 75 ans au service de l'heure universelle.* Edited by Martine M. Feissel and Suzanne Débarbat. Meudon, France: Observatoire de Paris-Bureau des Longitudes, 1992.

Parry, John. Letter to Joseph Renaud, 19 March 1917. In United Kingdom Hydrographic Office Archives, "Time Keeping at Sea."

———. "Opening Statement by the Chairman." [21 June 1917]. In United Kingdom Hydrographic Office Archives, "Time Keeping at Sea."

———. "History of the Inception of the I.H.B." *Hydrographic Review* 1 (1923): 13, 19.

Pasquier, Ernest. "De l'unification des heures dans la service des chemins de fer." 1889. Extract in *Ciel et Terre* 11 (1 August 1889): 266–70.

———. "Le 'temps universal' dans le système des fuseaux horaires." *Ciel et Terre* 11 (June 1890): 97–111.

———. "La Belgique et l'heure de Greenwich." 16 December 1890. *Ciel et Terre* 11 (January 1891): 497–509.

———. "L'unification de l'heure." Summarized in *Ciel et Terre* 12 (1 December 1891): 446–50, with extract, in English, in *Journal of the British Astronomical Association* 2 (November 1891): 106–7.

Pastorin, Don Juan. "Remarks on a Universal Prime Meridian." 30 April 1881. Translation in Fleming, *Universal or Cosmic Time,* suppl. papers, 49–51.

Pawson, E. "Local Times and Standard Time in New Zealand." *Journal of Historical Geography* 18 (1992): 278–87.

Philosophical Society of Washington. "Resolutions of the International Geodetic Association." *Bulletin of the Philosophical Society of Washington* 6 (1884): 107–9.

Ploix, Charles. "Utilié d'une entente entre les puissances maritimes pour l'adoption des mêmes éléments dans la publication des cartes et des instructions nautiques." In International Geographical Congress, IGC 2, 1:60–66.

Poincaré, Henri. "Rapport sur la proposition d'unification des jours astronomique et civil." In Bureau des Longitudes, *Annuaire du Bureau des Longitude pour 1895,* E.1–10. Translated and printed in *Observatory* 40 (September 1917): 323–27.

———. "Sur les signaux horaire destinés aux marins." *Comptes rendus* 150 (6 June 1910): 1471–72.

Renaud, Joseph. "L'heure à bord des navires." *Comptes rendus* 164 (29 January 1917): 221–22.

———. "L'heure en mer." In Bureau des Longitudes, *Annuaire pour l'an 1918*, B.1–45. Translated as "Memorandum by Monsieur Renaud" in United Kingdom Hydrographic Office Archives, "Time Keeping at Sea."

———. Letter to Admiral John Parry, 29 January 1917. Translated in United Kingdom Hydrographic Office Archives, "Time Keeping at Sea."

———. "La carte marine internationale." *Annales hydrographiques*, 3e sér., 2 (1918): 29–43.

———. "Le Planisphère des fuseaux horaires." *L'Astronomie* 32 (1918): 135–38, with chart.

Revue d'Astronomie Populaire, 1882–88.

Royal Astronomical Society. *Minutes of Council* 8 (14 November 1879): 49.

Royal Astronomical Society, Monthly Notices of, 1827–1910.

———. "The Proposed Change in the Astronomical Day. Letter from the Secretary to [*sic*; of] the Admiralty." *Monthly Notices of the Royal Astronomical Society* 78 (1918): 544–47.

———. "Commencement of the Astronomical Day: Letter from the Admiralty." *Monthly Notices of the Royal Astronomical Society* 79 (1919): 318.

Royal Geographical Society. "Sir J. H. Lefroy's Report on Mr. Sandford Fleming's Proposals Respecting a Prime Meridian for Time Reckoning." 19 November 1879. R.G.S. Archives.

———. "Standard Time at Sea." *Geographical Journal* 51 (February 1918): 97–100.

Royal Society. "Report," 6 November 1879. In Fleming, *Universal or Cosmic Time*, suppl. papers, 39.

Royal Society of Canada. Council. "Unification of Time." *Proceedings and Transactions of the Royal Society of Canada*, 2d ser., 4 (1898): xiv.

Royal Society of Canada, Proceedings and Transactions of, 1895–98.

———. "The Unification of Time at Sea." *Proceedings and Transactions of the Royal Society of Canada*, 2d ser., 2 (1896): Appendix A: A1–49.

———. "Unification of Time at Sea." *Proceedings and Transactions of the Royal Society of Canada*, 2d ser., 3 (1897): xxvii–xxx.

Sadler, Donald H. "Mean Solar Time on the Meridian of Greenwich." *Quarterly Journal of the Royal Astronomical Society* 19 (1978): 290–309.

Sadler, D. H., and G. A. Wilkens. "Astronomical Background to the International Meridian Conference of 1884." *Journal of Navigation* 38 (1985): 191–99.

Schram, Robert. "Einheitliche Zeit" (Uniform Time). *Wiener Zeitung*, 8 June 1886, 2–4; and 9 June 1886, 2–4.

———. "Zur Frage der Eisenbahnzeit" (On the Railroad Time Question). Pamphlet. From *Wiener Zeitung*, 14 December 1888, 2–4; and 15 December 1888, 2–4, 6.

———. "The Actual State of the Standard Time Question." *Observatory* 13 (April 1890): 139–46.

Science, 1883–86.

———. "International Standard Time." *Science* 1 (16 March 1883): 159.

———. "Resolutions of the International Geodetic Commission [Association]." *Science* 2 (28 December 1883): 814–15.

———. "Comment and Criticism." *Science* 3 (4 January 1884): 1.

———. "Unification of Time." *Science* 3 (25 April 1884): 517–18.

———. "Comment and Criticism." *Science* 3 (27 June 1884): 773–74.

———. "The Meridian Conference." *Science* 4 (1884): 376–78, 406, 421.

———. "Civil and Astronomical Time." *Science* 5 (17 April 1885): 308–9.

Smith, Humphry M. "Greenwich Time and the Prime Meridian." *Vistas in Astronomy* 20 (1976): 219–29.

Société de géographie, Bulletin de la, 1888–95.

Société royale belge de géographie, Bulletin de la, 1889–93.

Stephens, Carlene E. *Inventing Standard Time.* Washington: National Museum of American History, 1983.

Struve, F. G. W. *Expédition Chronométrique . . . entre Poulkova et Altona pour la détermination de la longitude géographique relative de L'Observatoire central de Russie.* St. Petersburg, 1846.

———. *Arc du méridien de 25°20′ entre le Danube et la Mer Glaciale.* St. Petersburg, 1857–60.

Struve, F. G. W., and Otto W. Struve. *Expédition Chronométrique . . . entre Altona et Greenwich pour la détermination de la longitude géographique de L'Observatoire central de Russie.* St. Petersburg, 1848.

Struve, Otto W. "Du premier méridien." *Bulletin de la Société de Géographie,* 6th ser., 9 (1875): 46–64. Translation of the original Russian text.

———. "On Universal Time and on the Choice for That Purpose, of a Prime Meridian." Translation of a report made to the Imperial Academy of Sciences on 30 September 1880. In *Proceedings of the American Metrological Society* 2 (December 1880): 170.

———. *Die Beschlüsse der Washingtoner Meridianconferenze.* St. Petersburg, 1885. English translation in Fleming, *Universal or Cosmic Time,* suppl. papers, 84–101.

Thrower, Norman J. W. *Maps & Civilization: Cartography in Culture and Society.* 2d ed. Chicago: University of Chicago Press, 1999.

United Kingdom. Admiralty. Letter from Secretary to Secretary, Science and Art Department, 5 January 1886. In United Kingdom, Science and Art Department, *Thirty-third Report* [for 1885], Appendix A, 38–39.

———. Letter from Secretary to Undersecretary of State, Colonial Office, 12 December 1895. In United Kingdom, Science and Art Department, *Forty-third Report* [for 1895], Appendix A, 15.

———. Letter from Secretary to Astronomer Royal Wm. H. M. Christie, 19 January 1897. In the Royal Greenwich Observatory Archives, 7/146.

———. Admiralty Order no. 1291. 9 April 1919. Noted in *Report of the Astronomer Royal to the Board of Visitors of the Royal Observatory, Greenwich,* 14 June 1919, 17–18.

————. Hydrographic Department. "The World Time Zone Chart." 21 March 1919. Identified as "D6 AA no.1, under the superintendence of J. F. Parry, Hydrographer."

————. "System of Time-keeping at Sea by means of Time Zones." 8 June 1920. *Admiralty Notices to Mariners*, no. 918.

————. "Method of Expressing Astronomical Time in Nautical Almanac; Intended Change." 28 June 1920. *Admiralty Notices to Mariners*, no. 1030.

————. Hydrographic Office. Series no. 5006, "Time Zone Chart of the World." First published 25 May 1921.

United Kingdom. Board of Trade. Marine Division. "Memorandum on Time Keeping at Sea," with "Time Zone Chart," X.301, 4 April 1918.

United Kingdom. Colonial Office. Secretary of State for the Colonies to Governor-General of Canada, 15 October 1879. Cited in Fleming, *Universal or Cosmic Time*, 31–32.

United Kingdom. *Public General Statutes*, 43 & 44 Vict., c.9.

United Kingdom. Science and Art Department. *[Yearly] Report of the Science and Art Department of the Committee of Council for Education*, 1882–98. London: Eyre & Spottiswoode.

United Kingdom. United Kingdom Hydrographic Office. Archives. Joseph Renaud to John Parry, 22 January 1917, from a translation in "Time Keeping at Sea," File no. Misc. 50, Folder no. 1.

————. "Time Keeping at Sea." [1917–18]. File no. Misc. 50, Folder no. 1.

————. "Conference on Time-Keeping at Sea." [April 1918]. 4 pp. In "Time Keeping at Sea," File no. Misc. 50, Folder no. 1.

United Kingdom. War Office. *Notes of the Government Surveys of the Principal Countries*. . . . London: Her Majesty's Stationery Office, 1882.

————. *A Guide to Recent Large Scale Maps*. . . . London: Her Majesty's Stationery Office, 1899.

U.S. Congress. House. *American Prime Meridian*, 31st Cong., 1st sess., 2 May 1850. H. Rept. 286.

U.S. Congress. Senate. Committee on Foreign Relations. *Concurrent Resolution Authorizing the President to Communicate to the Governments of All Nations the Resolutions Adopted by the International [Meridian] Conference . . . for Fixing a Prime Meridian*. . . . 48th Congr., 2d sess., 7 February 1885. S. Misc. Doc. 38.

————. *Report to Accompany Concurrent Resolution*. 48th Congr., 2d sess., 7 February 1885. S. Rept. 1188. Also *Congressional Record*, Senate, 9 February 1885, 16:1449–50.

————. *Letter from the Secretary of the Navy Recommending that the Meridian of Greenwich Be Adopted*. . . . 48th Congr., 2d sess., 13 February 1885. S. Ex. Doc. 73.

————. *Letter from the Secretary of the Navy Transmitting Communications Concerning the Proposed Change in the Time for Beginning the Astronomical Day*. 48th Congr., 2d sess., 18 February 1885. S. Ex. Doc. 78.

————. Committee on Naval Affairs. *Letter from the Secretary of the Navy Transmitting . . . Report of the National Academy of Sciences upon the Pro-*

posed New Naval Observatory. 49th Congr., 1st sess., 10 February 1886. S. Ex. Doc. 67.

U.S. Department of Commerce. National Bureau of Standards. *The International Bureau of Weights and Measures, 1875–1975.* Special Publication 420. Washington: GPO, 1975.

U.S. Department of the Navy. "Use of Standard Time Zones at Sea." *General Order no. 521.* 25 March 1920.

————. *Annual Reports of the Navy Department for the Year 1904.* Washington, 1904.

————. *Annual Reports . . . for the Year 1905.* Washington, 1906.

————. *Annual Reports . . . for the Fiscal Year 1912.* Appendix 2, "Report of the Superintendent of Naval Observatory," 180–81. Washington, 1913.

————. Secretary to president pro tempore of the Senate, 12 February 1885. In U.S. Senate, Committee on Foreign Relations, *Letter from the Secretary of the Navy . . . that the Meridian of Greenwich Be Adopted . . .* , 48th Congr., 2d sess., 13 February 1885. S. Ex. Doc. 73.

————. Secretary to president pro tempore of the Senate, 17 February 1885. In U.S. Senate, Committee on Foreign Relations, *Letter from the Secretary of the Navy Transmitting Communications Concerning the Proposed Change in the Time for Beginning the Astronomical Day,* 48th Congr., 2d sess., 18 February 1885. S. Ex. Doc. 78.

————. Secretary to O. C. Marsh, president of the National Academy of Sciences, 22 April 1885. In U.S. Senate, Committee on Naval Affairs, *Letter from the Secretary of the Navy Transmitting . . . Report of the National Academy of Sciences upon the Proposed New Naval Observatory,* 49th Congr., 1st sess., 10 February 1886. S. Ex. Doc. 67, Appendix: Document no. 1, 20–21.

————. Hydrographic Office. "U.S. Naval Wireless Telegraph Services." 22 November 1904. *Special Notice to Mariners,* no. 47a (1681).

————. Chart series H.O. 5192. First published May 1920; now published periodically by the National Geospatial-Intelligence Agency as "Standard Time Zone Chart of the World."

————. "Time Zones." *Notice to Mariners* 28 (10 July 1920): 2458.

————. Naval Observatory. [Regarding Resolution VI of the International Meridian Conference, Changing the Start of the Astronomical Day]. March–April 1886. In United Kingdom, Science and Art Department, *Thirty-fourth Report* [for 1886], Appendix A, 12–19.

————. Report Relative to the Astronomical Day. 16 October 1894. NARA, Records of the Department of State. RG59, Miscellaneous Letters, M179, roll 900.

U.S. Department of State. Secretary. Circular Letter, 23 October 1882. In U.S. President, *Message . . . Relative to International Meridian Conference,* 4 December 1884, Annex III.

————. Circular Letter, 1 December 1883. In U.S. President, *Message . . . Relative to International Meridian Conference,* 4 December 1884, Annex IV.

———. Secretary to F. A. P. Barnard, 17 September 1884. NARA, Records of the Department of State. RG59, Domestic Letters, M40, roll 101, ledger 433–34.

———. Secretary to Meridian Conference Delegates, 24 September 1884. NARA, Records of the Department of State. RG59, Domestic Letters, M40, roll 101, ledger 497–98.

———. Secretary to Chairman, Committee on Foreign Relations, 5 February 1885. In U.S. Senate, *Report to accompany Concurrent Resolution*, 48th Congr., 2d sess., 7 February 1885. S. Rept. 1188, 2.

———. United Kingdom Inquiry Letter Regarding the Astronomical Day, 29 September 1894. NARA, Records of the Department of State. RG59, Notes from the British Legation, M50, roll 124.

———. Report of the U.S. Naval Observatory, 16 October 1894, and cover letter, Secretary of the Navy, 20 October 1894. NARA, Records of the Department of State. RG59, Miscellaneous Letters, M179, roll 900, 16–31 October 1894.

———. Response to the United Kingdom Regarding the Astronomical Day. NARA, Records of the Department of State. RG59, Notes to the British Legation, M99, roll 51, 24 October 1894.

———. Letter to Cleveland Abbe, 8 January 1895. NARA, Records of the Department of State. RG59, Domestic Letters, M40, roll 125.

U.S. President. *Annual Message of the President*, 48th Congr., 2d sess., 1 December 1884. Ex. Doc. 1, viii.

———. *Message from the President of the United States Transmitting a Communication from the Secretary of State, Relative to International Meridian Conference Held at Washington*, 48th Congr., 2d sess., 4 December 1884. H. Ex. Doc. 14.

———. *Message from the President of the United States Transmitting a Report from the Secretary of State, Recommending the Government to Take Action to Approve the Resolutions of the International Meridian Conference Relating to Fixing a Prime Meridian and a Universal Day*, 50th Congr., 1st sess., 9 January 1888. H. Ex. Doc. 61.

Wayman, P. A. *Dunsink Observatory, 1785–1985*. Dublin: Royal Dublin Society and Dublin Institute for Advanced Studies, 1987.

Wheeler, G. M. *Report upon the Third International Geographical Congress and Exhibition at Venice, Italy, 1881*. 48th Congr., 2d sess., 1885. H. Ex. Doc. 270.

Wilkens, G. A. "The History of H.M. Nautical Almanac Office." In *Proceedings Nautical Almanac Office Sesquicentennial Symposium*. Washington: U.S. Naval Observatory, 1999, 55–81.

Willis, Bailey. "The International Millionth Map of the World." *National Geographic Magazine* 21 (February 1910): 125–32.

W[inlock], W. C. "The Proposed Change in the Astronomical Day." *Science* 5 (26 June 1885): 524–25.

Wisely, W. H. *The American Civil Engineer, 1852–1974*. New York: American Society of Civil Engineers, [1974].

X. Sentrick. "Time for the Continent." *Railroad Gazette* 14 (7 May 1870): 121.

III. TIME AS A SOCIAL INSTRUMENT (1883–1927)

A. P. "L'économie d'énergie électrique et l'avance de l'heure légale." *Le Génie Civil* 70 (3 February 1917): 81.

Bartky, Ian R. *Selling the True Time: Nineteenth-Century Timekeeping in America*. Stanford: Stanford University Press, 2000.

[Bartky, Ian R., James Filliben, Harry Ku, and Hans Oser]. *Review and Technical Evaluation of the DOT Daylight Saving Time Study*, NBS Report prepared for the Chairman, Subcommittee on Transportation and Commerce, Committee on Interstate and Foreign Commerce, U.S. House of Representatives, June 1976. Printed in Committee on Interstate and Foreign Commerce, *Daylight Savings [sic] Time Act of 1976: Hearings on H.R. 13089, H.R. 13090, and others*, 94th Congr., 2d sess., June 1976, 125–351. Serial 94-109.

Bartky, Ian R., and Elizabeth Harrison. "Standard and Daylight-saving Time." *Scientific American* 240 (May 1979): 36–43.

B[iggs], M. S. "Willett, William." *The Dictionary of National Biography, 1912–1921*. Oxford: Oxford University Press, 1927, 578.

Boston Chamber of Commerce. *Current Affairs*, 5 June 1916–29 April 1918.

———. Special Committee on [a] Daylight Saving Plan. "An Hour of Light for an Hour of Night." *Current Affairs* 7 (19 March 1917): Supplement.

———. "Chamber in Favor of Daylight Saving as a War Measure." *Current Affairs* 7 (30 April 1917): 1.

Chamber of Commerce of the United States. Committee on Daylight Saving. *Report of Committee on Daylight Saving of the Chamber of Commerce of the United States*. Washington, [February 1917].

Chicago Association of Commerce. "A Longer Sunlight Day." *Chicago Commerce* 10 (11 December 1914): 10–16.

———. "Report of Special Committee on Proposed Change of Time." 11 December 1914. In U.S. Senate. Committee on Interstate Commerce. *Standardization of Time: Hearing on S. J. Res. 135*, 64th Congr., 1st sess., 26 May 1916, printed in ibid. *Daylight Saving Time and Standard Time for the United States: Hearings on S. 1854*, 65th Congr., 1st sess., 3 and 10 May 1917, 47.

———. "Problem of Change of Time." *Chicago Commerce* 11 (23 July 1915): 41–43.

Churchill, Winston. "A Silent Toast to William Willett." In *The Collected Essays of Sir Winston Churchill*, edited by Michael Wolff. Vol. 3. [London]: Library of Imperial History, 1976.

de Carle, William. *British Time*. London: Crosby Lockwood & Son, [1947].

Doane, Doris Chase. *Time Changes in the U.S.A.* Downey, Calif.: Graphic Arts Press, 1966, with 1969 supplement tipped in.

Downing, Michael. *Spring Forward: The Annual Madness of Daylight Saving Time*. Washington: Shoemaker & Hoard, 2005.

Eyle, Alexandra. *Charles Lathrop Pack: Timberman, Forest Conservationist, and Pioneer in Forest Education.* Syracuse: ESP College Foundation, College of Environmental Science and Forestry, State University of New York, 1992.

[Franklin, Benjamin]. "A Subscriber to [the Journal's] Authors" (in French). *Journal de Paris* no. 116 (26 April 1784): 2–3.

———. "An Economical Project." In *The Writings of Benjamin Franklin,* edited by Albert H. Smyth. Vol. 9. New York: Macmillan, 1906.

Garland, Robert. *Ten Years of Daylight Saving from the Pittsburgh Standpoint.* Pittsburgh: Carnegie Library of Pittsburgh, 1927.

Hassinger, Hugo. "Für oder gegen die Sommerzeit?" *Kartographische Zeitschrift* 6 (January 1917): 11–12.

Herbert, Alan Patrick. *In the Dark: The Summer Time Story and the Painless Plan.* London: Bodley Head, 1970.

Horological Journal, 1907–18.

Howse, Derek. *Greenwich Time and the Longitude.* London: Philip Wilson, 1997.

Hudson, G. V. "On Seasonal Time." *Transactions and Proceedings of the New Zealand Institute* 31 (1898): 577–83, 719.

La Nature, 1910–19.

———. "Modification de l'heure légale." *La Nature* 44 (13 May 1916): 319.

Lallemand, Charles. "Sur un projet de modification de l'heure légale." *Comptes rendus* 162 (10 April 1916): 536–42.

Le Génie Civil, 1916–19.

———. "Le projet de modification de l'heure légale en France." *Le Génie Civil* 68 (22 April 1916): 269–70.

Nature, 1908–18.

———. "The Daylight Saving Bill." *Nature* 78 (9 July 1908): 223–26.

New York Times, 1907–18, 1934–37.

O'Malley, Michael. *Keeping Watch: A History of American Time.* New York: Viking Penguin, 1990.

P. D. "Les économies de combustible et l'avance de l'heure légale." *Le Génie Civil* 69 (30 December 1916): 443–44.

Pack, Charles Lathrop. *War Gardens Victorious.* Philadelphia: Press of J. B. Lippincott, 1919.

Pollak, Oliver B. "'Efficiency, Preparedness and Conservation': The Daylight Savings Time Movement." *History Today* 31 (March 1981): 5–9.

Popular Astronomy. "Detroit's Different Kinds of Time." *Popular Astronomy* 9 (1901): 36–37.

Prerau, David. *Seize the Daylight: The Curious and Contentious Story of Daylight Saving Time.* New York: Thunder's Mouth Press, 2005.

Renaud, Joseph. "L'avance de l'heure légale pendant l'été de l'année 1916." In Bureau des Longitudes, *Annuaire pour l'an 1917,* B.37.

Times (London), 1908–19, 1925–27.

Turner, E. S. "Taking Time by the Forelock." In *Roads to Ruin: The Shocking History of Social Reform.* London: Michael Joseph, 1950.

Turner, H. H. "Daylight Saving." *Observatory* 39 (October 1916): 419–25.

United Kingdom. Home Department. Summer Time Committee. *Report of the*

Committee Appointed by the Secretary of State for the Home Department to Enquire into the Social and Economic Results of the Summer Time Act, 1916. . . . London: His Majesty's Stationery Office, 22 February 1917.

United Kingdom. Home Office and Scottish Home and Health Department. *Review of British Standard Time.* London: Her Majesty's Stationery Office, 1970.

United Kingdom. Parliament. House of Commons. *Report, and Special Report, from the Select Committee on the [1908] Daylight Saving Bill, Together with the Proceedings of the Committee, Minutes of Evidence, and Appendix.* London: Printed for His Majesty's Stationery Office by Vacher and Sons, 1908.

————. *Report and Special Report from the Select Committee on the [1909] Daylight Saving Bill; together with the Proceedings of the Committee, Minutes of Evidence, and Appendix.* London: Printed for His Majesty's Stationery Office by Wyman and Sons, 1909.

U.S. Congress. *Congressional Record*, 1909, 1917–19.

U.S. Congress. House. Committee on Interstate and Foreign Commerce. *Daylight Saving,* 65th Congress, 2d sess., 9 February 1918. H. Rept. 293.

————. *Hearings on Repeal of Section Three of the Daylight Saving Act: H.R. 3854,* 66th Congr., 1st sess., 13 June 1919.

————. *Hearings on Repeal of Daylight Saving Law,* 66th Congr., 1st sess., 13 June 1919. H. Rept. 42.

————. *Daylight Savings [sic] Time Act of 1976: Hearings on H.R. 13089, H.R. 13090, and others,* 94th Congr., 2d sess., June 1976. Serial 94-109.

U.S. Congress. Senate. Committee on Interstate Commerce. *Standardization of Time: Hearing on S. J. Res. 135,* 64th Congr., 1st sess., 26 May 1916. Printed in *Daylight Saving Time and Standard Time for the United States: Hearings on S. 1854,* 65th Congr., 1st sess., 10 May 1917, 43–48.

————. *Daylight Saving Time and Standard Time for the United States: Hearings on S. 1854,* 65th Congr., 1st sess., 3 and 10 May 1917.

————. *Daylight Saving,* 65th Congr., 1st sess., 25 May 1917. S. Rept. 46.

U.S. Department of Commerce. National Bureau of Standards. *NBS Report,* 1976, as printed in U.S. Congress, House, Committee on Interstate and Foreign Commerce, Serial 94-109.

U.S. Department of Transportation. Office of the Secretary. *Standard Time in the United States: A History of Standard and Daylight Saving Time in the United States and an Analysis of the Related Laws.* Washington, July 1970.

————. Lawson, Linda L., Acting Deputy Assistant Secretary for Transportation Policy. "Prepared Statement." In U.S. Congress, House, Science Committee, *Energy Conservation Potential of Extended and Double Daylight Saving Time: Hearing before the Energy Subcommittee,* 107th Congr., 1st sess., 24 May 2001, 17–20. Serial 107-30.

U.S. Interstate Commerce Commission. "No. 10122: Standard Time Zone Investigation." *I.C.C.* 51, no. 273 (24 October 1918): 275–77.

U.S. Supreme Court. Massachusetts State Grange v. [Attorney General] Benton. *United States Supreme Court Reports.* 272 U.S. 525 (1926).

Waymark, Peter. *A History of Petts Wood*. 3d ed. N.p.: Petts Wood and District Residents Association, 1990.

Willett, William. *The Waste of Daylight*. [London]: n.p., 1907.

———. "The Waste of Daylight." *Horological Journal* 50 (September 1907): 4–7.

———. *The Waste of Daylight: Extracts from Letters and Newspapers with Names of Supporters of Mr. W. Willett's Proposal for Legislative Action*. Proof copy. [London: Spottiswoode, January 1908].

———. *The Waste of Daylight*. London: Spottiswoode, [April] 1908.

———. Letter-to-the-Editor. *Horological Journal* 50 (April 1908): 144.

———. *Le gaspillage de la lumière du jour*. [Paris (?), 1910 (?)].

———. *The Waste of Daylight: Opinions of Eminent Men, Text of Bill . . . , Statement of Progress. . . .* [London]: n.p., March 1911.

———. *The Waste of Daylight: With an Account of the Progress of the Daylight Saving Bill*. 12th ed. [London]: N.p., March 1912.

———. *The Waste of Daylight: With an Account of the Progress of the Daylight Saving Bill*. 19th ed. [London]: N.p., March 1914.

———. Letter. *Times* (London), 17 April 1914, 11.

Wright, Thomas D. Letter. Reprinted in *Horological Journal* 50 (1908): 23–24.

Index